W9-ACX-247

BAMBOOS OF CHINA

BAMBOOS OF CHINA

Wang Dajun & Shen Shao-Jin

TIMBER PRESS *Portland, Oregon*

Drawings and paintings: Lian Gang
Color photographs: Fong Yong-Xi
Black and white photographs: Fong Yong-Xi and Wang Dajun
Cultural contributions: Ni Kan

ISBN 0-88192-074-6

Printed in Hong Kong
Designed by Sandra Mattielli

TIMBER PRESS
9999 SW Wilshire
Portland, Oregon 97225

CONTENTS

This is an ideal place for retirees. In ancient China, intellectuals worked as high-ranking officials in the royal court. They often dreamed about retiring. "Accompanying the Emperor is just like accompanying a tiger." This proverb explains how dangerous it was to be a court official.

1 BAMBOO AND CULTURE

Should one ask what plant, other than the main food crops, has had the most profound influence on the life and culture of the Chinese people, one could say with full confidence that it is the bamboo. In the prehistoric age, when people had to fight or to hunt for life, the arrow, the then most advanced and dreadful weapon, owed its shaft to a bamboo. In one of the six Sacred Books, *Shu Jing (The Book of History)*, it is recounted that under the reign of Emperor Yu (about 2000 B.C.) three small dukedoms near Yun Men Ze (Lake of the Dreaming Cloud, in the central Yangtze River Valley), were ordered to pay a yearly tribute of bamboos suitable for making arrow shafts.

Official documents, records and histories of ancient China were written on thin, narrow bamboo slips about 30 cm long, called *Jian.* Many slips strung together became a book. A scholar had to be strong to carry his library of bamboo slips in a large bale on his back. For three or four hundred years after the invention of the 'silk' book, bamboo slip books were still in circulation. When the first Emperor of the Qin Dynasty (Qin Shi Huang 259–210 B.C.) ordered the books of Confucianism (mainly the six Sacred Books) burned and a few dozen leading Confucian scholars buried alive, many copies of the forbidden books were preserved by burying them in the ground. They were unearthed in the Han Dynasty (207 B.C. to A.D. 220), long after the silk books had rotted while the bamboo slip books remained whole and readable. Several bamboo slip books, recently excavated from the tombs of ancient scholars, are still readable.

Chinese music owes a great deal to bamboo. A prehistoric legend tells of how the temperament of music was settled. Huang Ti (Yellow Emperor, about 2700 B.C.) ordered a court musician, Ling Lun, to establish an official standard for musical temperament. Ling made twelve bamboo pipes as the standard set: six mimicked the melodious singing of the female phoenix, while the remaining six the sonorous voice of the male. This myth is too sophisticated and beautiful to have been created by the simple culture of the ancients, and its genesis is now placed at least one thousand years later. In any event, many kinds of Chinese musical instruments are made of bamboo, such as the various types of popular two-stringed, bowed instruments, and several kinds of pipes and flutes.

More than four thousand years ago, bamboo had become one of the principal materials used in the everyday life of the Chinese people. Common housewares, such as baskets, cases, boxes, chairs, mats, handles of farm implements, etc., were made of bamboo. Numerous exquisite handcrafted articles were woven out of very thin and very narrow bamboo strips. It was a common practice to build bamboo houses at least two thousand years ago. Even the roof was made of large bamboo culms split in half lengthwise. A classical story, *The Two-Story Bamboo House at Huangkan* by Wang Yu-Cheng (A.D. 954–1001) in the Song Dynasty, informs us that bamboo houses would last at least twenty years without repairs. Pavilions and open galleries made of bamboo have decorated Chinese gardens for centuries.

The cultivation of bamboo became one of the most profitable business undertakings in early China according to the great historical work *Shi Ji (History)* by Si-ma Qian (145–? B.C.). The yearly income from a 1000 mu (about 67 ha) grove of bamboo in the Weihe River basin equaled the revenue from taxes collected by a marquis from one thousand households of farmers. About two thousand years ago, large groves of bamboo were grown commercially in the Quishui River basin (southeast of Mount Taihangshan) and the Weihe River basin (central Shansi Province). They are still the principal producing areas in Northern China today.

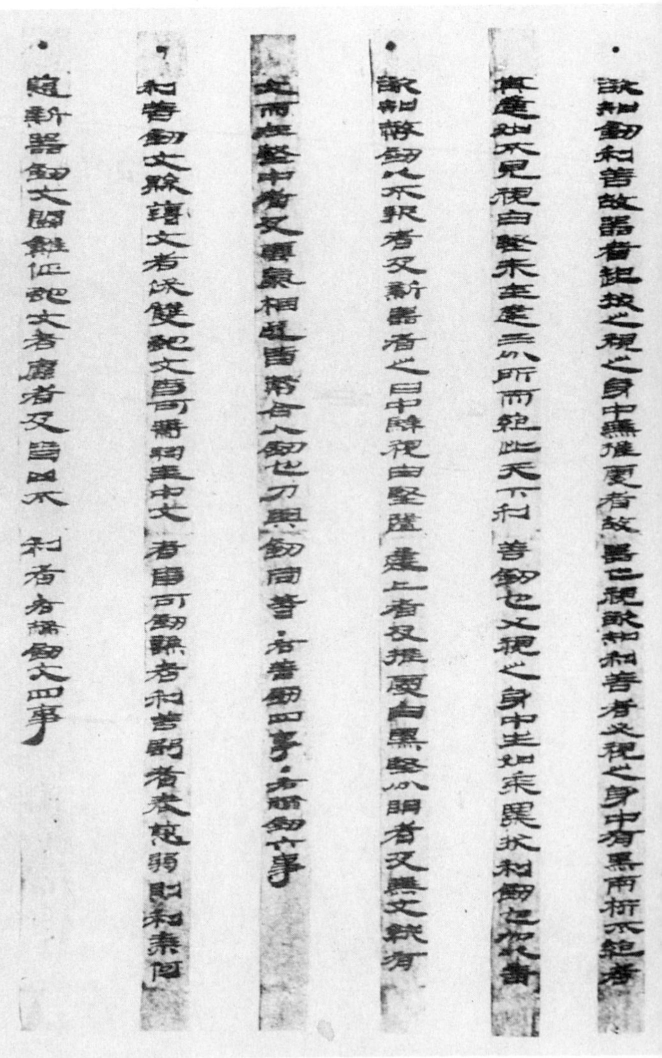

Bamboo slip book, 88 B.C. (Fong)

Bamboo slip book, Eastern Han Dynasty. (Fong)

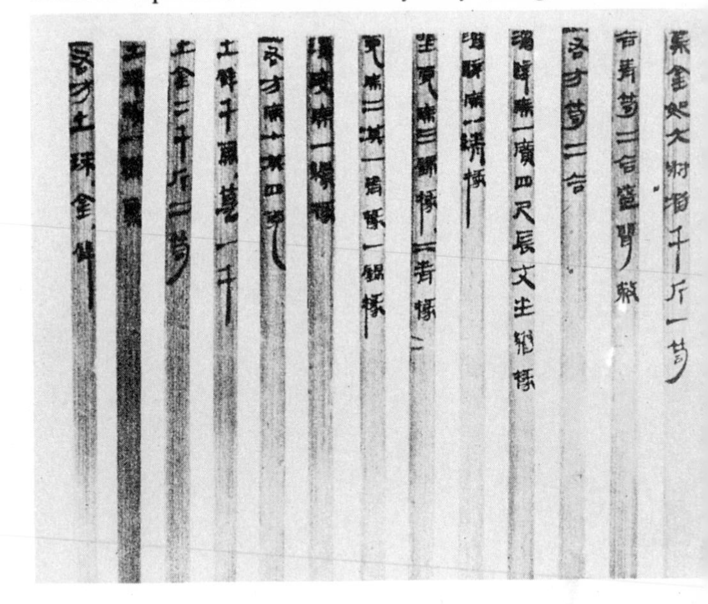

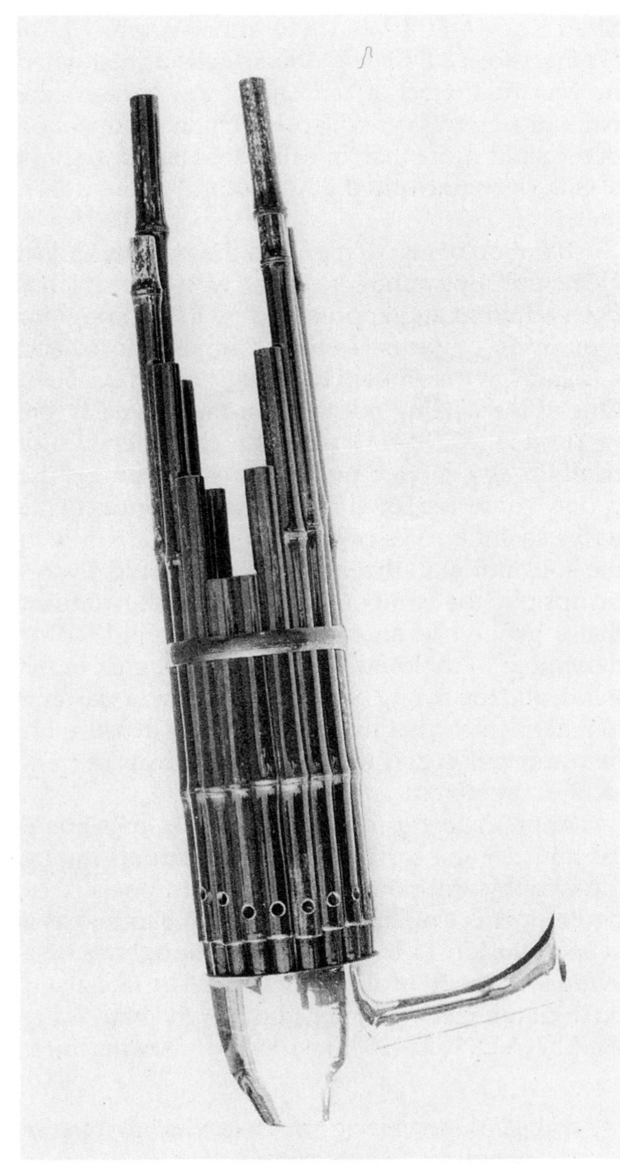

Sheng, a Chinese wind instrument made of bamboo. (Fong)

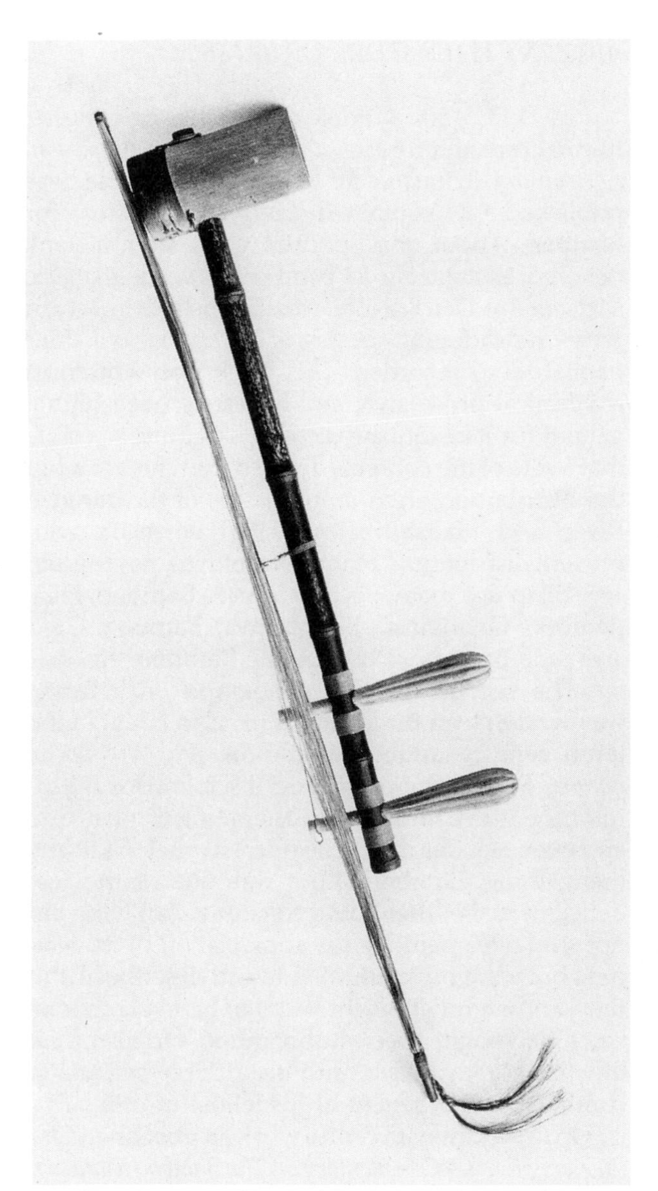

Jianghu, a 2-stringed bowed instrument made of bamboo. The most important instrument in the Beijing opera. (Fong)

Chinese bamboo pipes. (Fong)

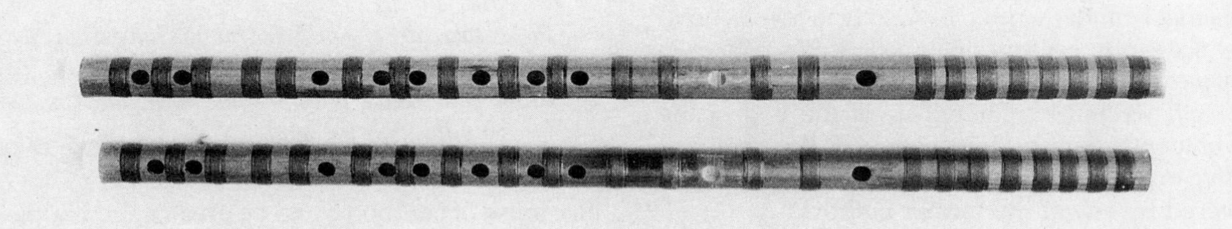

ANCIENT LITERATURE OF BAMBOO

In A.D. 450, a book dealing in agriculture, horticulture and forestry, *Qi Min Yao Shu (Important Agricultural Techniques for People)* by Jia Si-Zie, was published. It contained cultivation notes for bamboo. About one hundred years later, a book devoted exclusively to bamboo, *Zhu Pu (Bamboo Manual)* by Dai Kai-Zhi was published in which sixty-one different species of bamboos and their habitat were recorded. This book was written in rhythmical prose style and has since been highly valued for its exquisite diction. It is quite possible that some of the contents derived from hearsay, but the distribution areas in the basins of the Yangtze River and Hanshui River cited correlate with present distribution. Many common names therein are still in use today, such as Square Bamboo, Filial Bamboo, Guanyin (a Bodhisattva) Bamboo, Gold and Jade Bamboo, Phoenix-Tail Bamboo, etc.

The second known monograph on bamboo was written by a Buddhist monk, Zan Ning, in the tenth century under the title *Sun Pu (Manual of Bamboo Shoot)*. The prescribed discipline for Buddhist monks in China forbade the eating of meat, fish, and even eggs, as it was considered cruel to kill any living thing. Bamboo shoot was one of the few delicious dishes they could consume. Zan Ning did not stop after penning his appreciation of its tastiness but went on to enumerate and describe all the then-known ninety-eight kinds of bamboo shoots; i.e., ninety-eight species of bamboo. He also took the trouble to collect and record recipes using bamboo for the benefit of his fellow monks.

In the fourteenth century, Li Kan published *Zhu Pu Xiang Kao (A Commentary on The Bamboo Manual)*. In the same century Liu Mei wrote *Zhi's Shu Zhu Pu (Bamboo Manual Second)*, which added twenty-two species of previously unknown bamboos to the old manual of Zan Ning.

In the middle of the fifteenth century, an important reference book, Yu Zong-Ben's *Zhong Shu Shu (The Book of Arboriculture)* appeared. The book, highly esteemed for its practical and scientific value, devoted an entire chapter to the cultivation of bamboo. *Zun-Sheng Ba Jian (Zun-Sheng's Eight Commentarial Writings)* written by Gao Lian at the end of the sixteenth century included a new *Zhu Pu (Bamboo Manual)*. An entirely new *Zhu Pu (Bamboo Manual)* written by Chen Ting at the end of the seventeenth century dealt only with the bamboos native to Southwestern China, which had not been covered by any of the former authors.

There is even a record of fossil bamboo shoots in the old literature. The famous scientist and writer Shen Kuo (A.D. 1030–1095) in his *Men Xi Hi Tan (Writings from a Dreaming Stream)* tells us that when he was the Prefect of Yan Zhou (now Yan-an), the bank of a large river collapsed. Upon examination of the slide, more than one hundred bamboo shoot fossils were unearthed at a depth of a few dozen feet.

Bamboo being so useful to the people, and so elegant in appearance, it is small wonder that it has been admired and appreciated by the Chinese for thousands of years. Indeed Chinese poets and scholars have attributed to it near-human qualities. One of the leading poets of the Tang Dynasty, Bai Jy-Yi (A.D. 772–846), using the characteristics of bamboo as a model, wrote a prescription for the proper ordering of a gentleman's conduct. He wrote that the roots of bamboo, growing firmly in the soil, indicated that a gentleman should always be upright; the bamboo culm being hollow meant that a gentleman should never be prejudiced or dissemble by holding back secret thoughts in his mind; and the strong bamboo node gave a warning to a gentleman that he should always treasure his own reputation and never yield to threats or transient allurements.

In order to express their bitter feelings about the injustice and corruption at the court and among the wealthy and powerful, the ancient poets often wrote poems extolling the virtues of bamboo as a veiled allusion to the decent and upright people living in a period of abounding evil. The farsighted political reformer, premier, and literary man, Wang An-Shi (A.D. 1021–1086) wrote the following lines:

Ascending the winding path through bamboo groves brings the coolness of the hall.
Your slanting shadow and whistling sound linger long with meaning.
We admire your uprightness, yet are filled with compassion for your leanness.
Of your great wisdom becoming more resolute with age we are all aware.
You like the weeds, ask nothing from people but rain and dew.
You follow the example of conifers, defying the bitter frost and ice.
Please take care of your roots for you have a long way to go.
From your stems phoenix pipes beyond price will ring through the land.

No praise of bamboo could be greater and I hope it will give the reader a deep understanding of the feeling of Chinese people about this splendid plant.

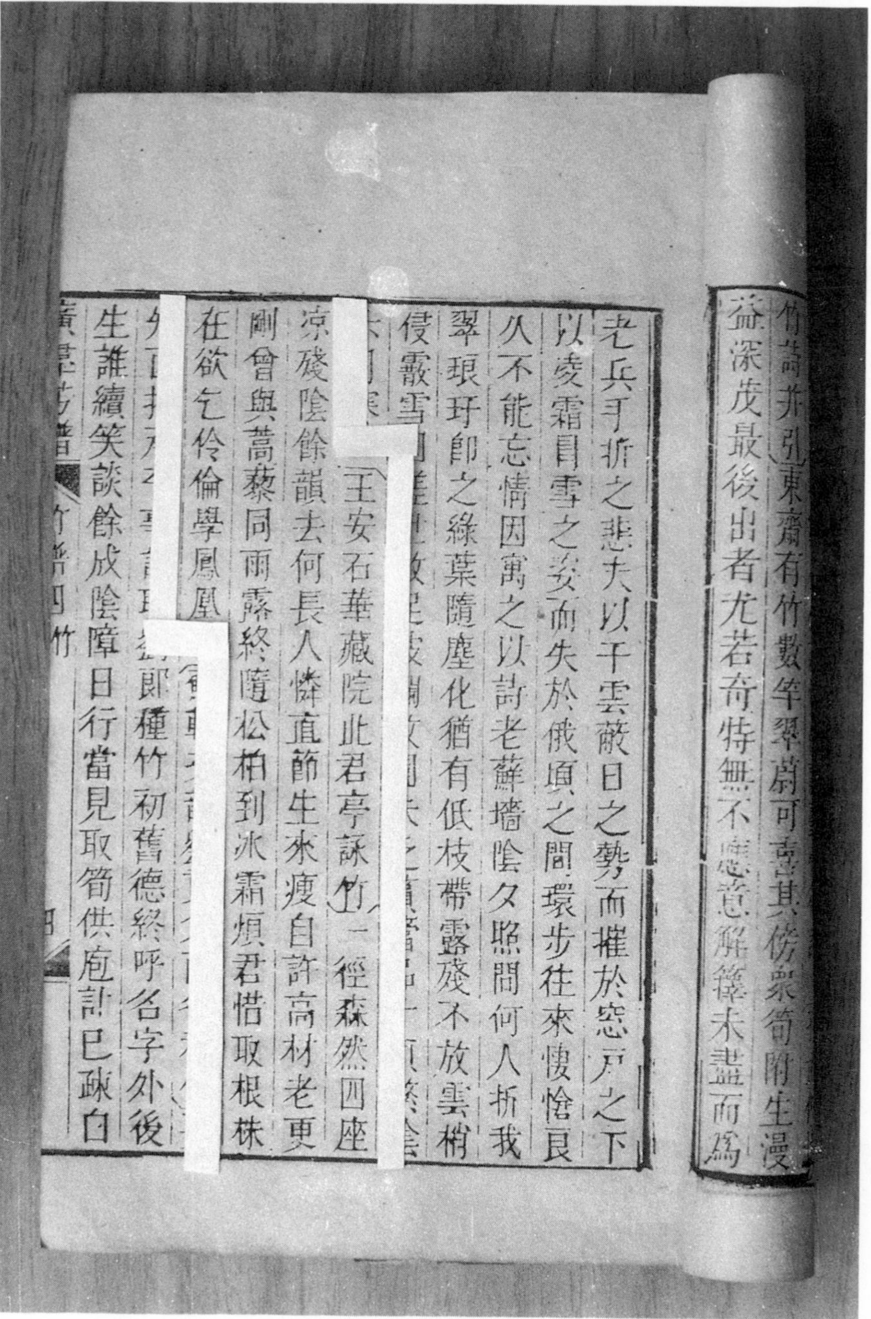
竹詩并引 東齋有竹數竿翠蔚可喜其傍眾筍附生漫
益深茂最後出者尤若奇特無不應意解籜未盡而為
老兵刃折之悲夫以干雲蔽日之勢而摧於窗戶之下
以凌霜冒雪之姿而失於俄頃之間環步往來悽愴良
久不能忘情因寓之以詩老蘚牆陰夕照間何人折我
翠琅玕郎之綠葉隨塵化猶有低枝帶露殘不放雲梢
侵霰雪[illegible]
[illegible]王安石華藏院此君亭詠竹一徑森然四座
涼殘陰餘韻去何長人憐直節生來瘦自許高材老更
剛曾與蒿藜同雨露終隨松柏到冰霜煩君惜取根株
在欲乞伶倫學鳳凰[illegible]
[illegible]
[illegible]郎種竹初舊德終呼名字外後
生誰續笑談餘成陰障日行當見取筍供庖計已疎白
廣群芳譜 竹譜四 竹

Wang An-Shi's poem.

癸亥孟秋铸剑楼晓钢

2 DISTRIBUTION OF BAMBOOS

Bamboos are long-lived, woody, evergreen grasses, members of the family *Gramineae,* tribe *Bambuseae.* The giant bamboos often reach a height of 20 to 30 m, while most of them are shrub-like medium or dwarf species with a few exceptions of climbing species. There are about 60 to 70 genera and over 1,200 species in the world. Most of them require a warm climate, abundant moisture, and fertile soil. Eastern and southeastern Asia are the center of its distribution. Only a few species can grow in fairly cold weather (below −20°C).

The adaptability of bamboos is admirable. You can find them growing in plains, hilly regions and high-altitude mountainous areas. They can grow in most kinds of soil except alkaline soils, dry desert, and marsh.

There are probably more than 30 genera and a few less than 300 species of bamboos native to China. Between 18°N to 35°N and 85°E to 122°E in China is the distribution area of all Chinese bamboos. But by acclimatizing and careful cultivating in the gardens of Beijing at 40°N, *Phyllostachys bambusoides, Ph. nigra,* and *Ph. propinqua* grow quite happily as ornamental plants. A native bamboo, *Pleioblastus chino,* hardy and dwarf, grows in the mountainous region of Liao Dong Peninsula, at approximately 39°N.

The range of vertical distribution is rather astonishing. At an altitude of 3,500 m in the Himalayas, at 2,300 m in the Qin Ling Mountains, or at 3,000 m in Xin Gao Shan of Taiwan, alpine bamboos flourish. But from altitudes 1,000 m up they are dwarf and shrub-like. The giant bamboos can only be found from sea level to 800 m.

According to the hardiness of bamboo species, there are three regions in China indicating the natural distribution of these important economical plants, which are also highly ornanmental.

North Region (between the Yellow River Valley and the North of the Yangtze River, mainly stoloniferous forms). From 30°N to 37°N. Yearly Mean Temperature: 12° to 17°C. Mean temperature in January: −4° to 4°C. Yearly precipitation: 500 to 1,200 mm.

In this region, the dominant stoloniferous species are *Phyllostachys bambusoides* and *Ph. nigra* var. *henonis;* but an intermediate type, having sympodial rhizomes with diageotropical growth and scattered culms, *Fargesia spathacea* is also important in the western mountainous areas.

Central Region (from the South of the Yangtze River to the north side of the Nan Ling Mountains). From 25°N to 30°N. Yearly Mean Temperature: 15°

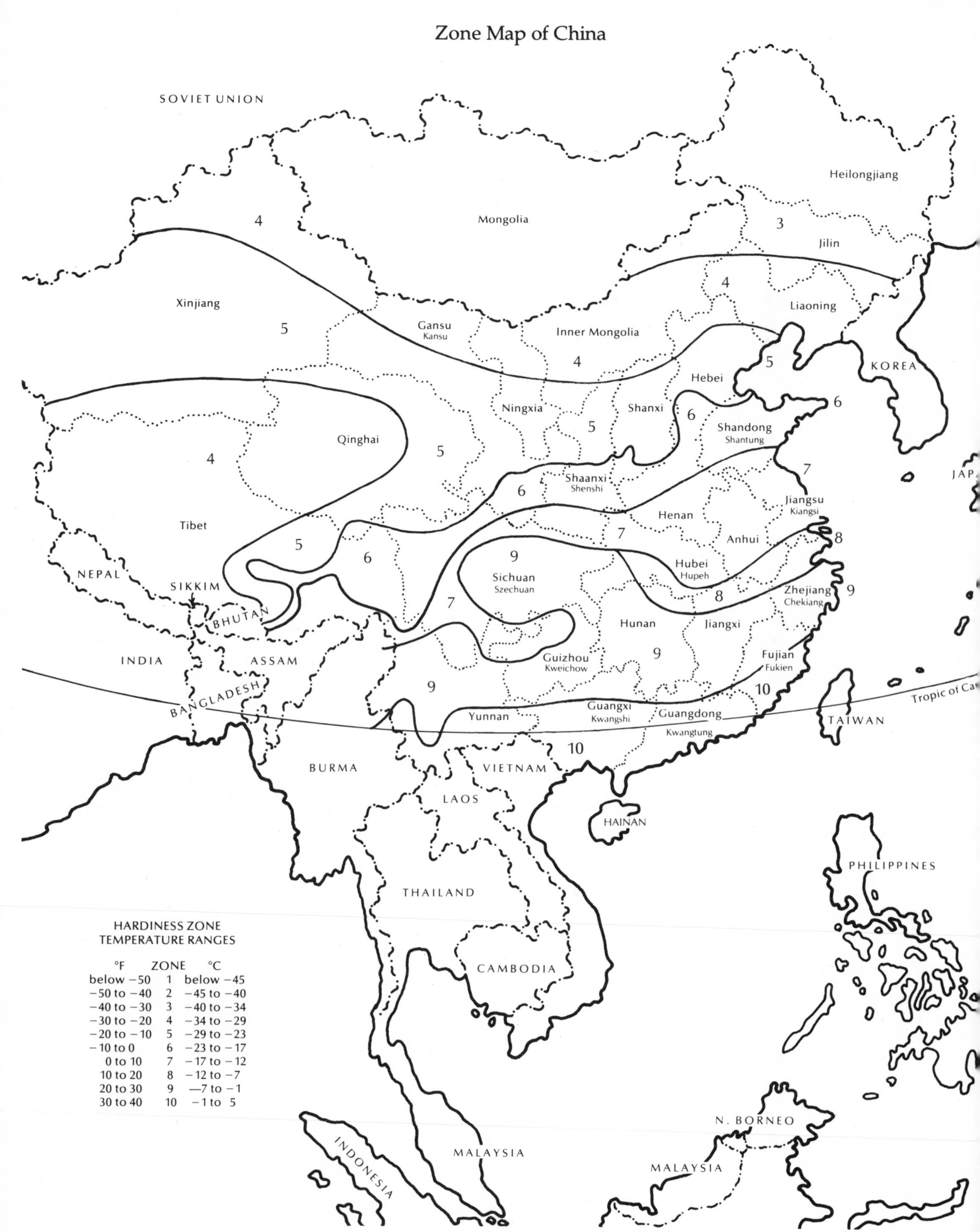
Zone Map of China
SOVIET UNION
Mongolia
Heilongjiang
Jilin
Liaoning
KOREA
Xinjiang
Gansu
Kansu
Inner Mongolia
Hebei
Ningxia
Shanxi
Shandong
Shantung
Qinghai
Shaanxi
Shenshi
Henan
Jiangsu
Kiangsi
Anhui
Tibet
Sichuan
Szechuan
Hubei
Hupeh
Zhejiang
Chekiang
Hunan
Jiangxi
Guizhou
Kweichow
Fujian
Fukien
Yunnan
Guangxi
Kwangshi
Guangdong
Kwangtung
TAIWAN
Tropic of Ca
JAP
NEPAL
SIKKIM
BHUTAN
INDIA
ASSAM
BANGLADESH
BURMA
VIETNAM
LAOS
HAINAN
THAILAND
CAMBODIA
PHILIPPINES
N. BORNEO
MALAYSIA
MALAYSIA
INDONESIA
3
4
4
4
4
5
5
5
5
5
5
6
6
6
6
7
7
7
8
8
9
9
9
9
10
10
HARDINESS ZONE
TEMPERATURE RANGES
°F ZONE °C
below −50 1 below −45
−50 to −40 2 −45 to −40
−40 to −30 3 −40 to −34
−30 to −20 4 −34 to −29
−20 to −10 5 −29 to −23
−10 to 0 6 −23 to −17
0 to 10 7 −17 to −12
10 to 20 8 −12 to −7
20 to 30 9 —7 to −1
30 to 40 10 −1 to 5

to 20°C. Mean temperature in January: 4° to 8°C. Yearly precipitation: 1,200 to 1,800 mm.

In the northern part of this region, the most important stoloniferous species are *Phyllostachys pubescens, Ph. bambusoides* var. *sulphurea, Ph. nigra, Ph. makinoi, Ph. congesta,* and *Ph. propinqua.* Among the intermediate type, *Pleioblastus amarus, Indocalamus latifolius,* and *I. tessellatus* are also widely distributed.

On the other hand, in the southern part of the region, clumped species are more common than stoloniferous. *Sinocalamus affinis, S. distegius, S. farinosus, Bambusa multiplex,* and *B. rigida* are common in this area.

South Region (from the South of the Nan Ling Mountains to Hainan Island, mainly clumped bamboos). From 25°N south. Yearly Mean Temperature: 20° to 22°C. Mean temperature in January: 8°C plus. Yearly precipitation: 1,200 to 1,800 mm (up to 3,000 mm in certain areas).

In high mountainous areas of the northern border of this region, both stoloniferous and intermediate type bamboos grow abundantly. But in other places, only clumped species are important, such as *Bambusa pervariabilis, B. rigida, B. textilis, B. sinospinosa, Sinocalamus latiflorus, S. oldhami, S. beecheyanus,* and its var. *pubescens, S. giganteus, Lingnania chungii, Schizostachyum funghomii,* etc.

GEOGRAPHIC DISTRIBUTION OF BAMBOO SPECIES

1. Liao-Tong Peninsula
Pleioblastus chino Makino, reported to be native to this area, can no longer be found.
The bamboos grown in gardens are *Phyllostachys* spp., which are also found in Beijing.

2. Beijing (Grown in Gardens)
Indocalamus latifolius, Phyllostachys aureosulcata, Ph. bambusoides, Ph. bambusoides var. *castilloni, Ph. bambusoides* var. *castilloni-inversa, Ph. glauca, Ph. nigra, Ph. nigra* var. *henonis, Ph. propinqua* are grown in sheltered places in gardens.

3. Weihe River Area
Phyllostachys glauca, Ph. viridis, and *Pleioblastus amarus* are the main spp. grown in this area.

4. Mountainous Areas
Alpine bamboos and *Arundinaria fangiana, Fargesia* spp., *Phyllostachys arcana, Ph. nigra* var. *henonis, Sinocalamus affinis* are common in this area.

5. East Yellow River Valley
Chimonobambusa quadrangularis, Phyllostachys bambusoides, Ph. decora, Ph. glauca, Ph. nigra, Ph. propinqua, Ph. viridis, Ph. vivax are commonly grown, while *Ph. pubescens* was recently introduced to this area.

6. Eastern Xizang
Chimonobambusa metuoensis, Fargesia spp., *Yushania* spp.

7. Yangtze River Valley
Bambusa multiplex, Indocalamus tessellatus, Phyllostachys aurea, Ph. bambusoides, Ph. heteroclada, Ph. glauca, Ph. iridenscens, Ph. nigra, Ph. praecox, Ph. propinqua, Ph. pubescens, Ph. viridis, Pleioblastus amarus, Shibataea chinensis, Sh. hispida are now popular in gardens.

7a. Eastern and Central Sichuan
More bamboo spp. are used as ornamentals, such as *Bambusa prominens, B. sinospinosa, Indocalamus pedalia, I. victorilis, I. emeiensis, Qiongzhuea opienensis, Q. rigidula.*

8. South China
Acidosasa chinensis, Ampelocalamus actinotrichus, Arundinaria amabilis, Bambusa spp., *Dendrocalamus strictus, Dinochloa* spp. (in Hainan Islands), *Indocalamus victorialis, Indosasa* spp., *Lingnania* spp., *Pseudosasa amabilis, Schizostachyum* spp., *Shibataea kumasasa, Sinobambusa* spp., Sinocalamus spp.
In southeastern Xizang *Bambusa lixin* and *Dendrocalamus tibeticus* are among the native plants.

9. Southern Yunnan
Bambusa sinospinosa, Chimonocalamus spp., *Dendrocalamus sinicus, D. strictus, Indosasa* spp., *Leptocanna chinensis, Qiongzhuea tumidinoda,* and *Sinocalamus giganteus* are distinguished spp.

SELECTED REFERENCES

Botanical Institute of Beijing. *Iconographia Cormophytorum Sinicorum (Vol. 5),* 1976. Beijing: Science Publishing Co.

Botanical Institute of Southern China. *Guangzhou Flora,* 1956. Beijing: Science Publishing Co.

Botanical Institute of Guangdong. *Hainan Flora (Vol. 4),* 1977. Beijing: Science Publishing Co.

Chao, Chi-son; Chu, Cheng-de. "A Study on the Bamboo Genus *Indosasa* of China." *Acta Phytotaxonomica Sinica* Vol. 21, No. 1. 1983. Beijing.

Chen, Shou-liang. *Gramineae in Eastern China.* 1962. Nanjing: Jiangsu People's Publishing Co.

Chen, Shou-liang; Chen, Shao-yun; Sheng, Guo-ying. "A Revision of Subtribe *Pleioblastinae* Keng & Keng f." *Acta Phytotaxonomica Sinica* Vol. 21, No. 4. 1983. Beijing.

Chia, Liang-chi; Fung, Hok-lam. "*Leptocanna,* A New Genus of *Bambusoideae* from China." *Acta Phytotaxonomica Sinica* Vol. 19, No. 2. 1981. Beijing.

Chia, Liang-chi; Fung, Hok-lam. "New Species of the Genus *Bambusa* Schreber from China." *Acta Phytotaxonomica Sinica* Vol 19, No. 3. 1981.

Chu, Cheng-de; Chao, Chi-son. "Acidosasa—A New Genus of *Bambusoideae* Native to China." *Journal of Nanjing Technological College of Forest Products.* 1979. Nanjing.

Chu, Cheng-de; Chao, Chi-son. "*Arundinaria* and Its Distribution in China." *Journal of Nanjing Technological College of Forest Products.* No. 3. 1980. Nanjing.

Chu, Cheng-de; Chao, Chi-son. "New Classification of *Bambusoideae* Native to China." *Journal of Nanjing Technological College of Forest Products.* No. 3. 1981. Nanjing.

Chu, Cheng-de; Chao, Chi-son. "New Spp. of *Bambusoideae* Native to Guizhou." *Journal of Bamboo Research* Vol. 1, No. 1. 1982. Nanjing.

Faculty of Biology, Nanjing University; Institute of Botany, Academia Sinica. *Pictorial Manual of Main Plants Native to China (Vol. Gramineae).* 1959. Beijing: Science Publishing Co.

Hsueh, Chi-ju; Yi, Tong-pei. "Two New Genera of *Bambusoideae* from S.W. China (1. *Chimonocalamus* Hsueh et Yi)." *Acta Botanica Yunnanica* Vol. 1, No. 2. 1979.

Hsueh, Chi-ju; Yi, Tong-pei. "Two New Genera of *Bambusoideae* from S.W. China (2. *Qiongzhuea* Hsueh et Yi)." *Acta Botanica Yunnanica.* Vol. 2, No. 1. 1980.

Hu, Cheng-hua. "A Taxonomical Study of the Genus *Sasamorpha* from China." *Journal of Bamboo Research* Vol. 2, No. 1. 1983. Hanzhou.

Keng, P. C. "A Revision of the Genera of Bamboos from the World." *Journal of Bamboo Research* Vol. 1, No. 1. 1982. Hanzhou.

Keng, P. C.; Xue, Ji-ru. "*Ferrocalamus* Hsueh et Keng f.—A New Bamboo Genus in China." *Journal of Bamboo Research* Vol. 1, No. 1. 1982. Hanzhou.

Nanjing Technological College of Forest Products. *The Culture of Bamboo Forest.* 1981. Beijing: Chinese Forestry Publishing Co.

The Editing Committee of Chinese Vegetation. *Chinese Vegetation.* 1980. Beijing: Science Publishing Co.

Wang, Zheng-ping; Chu, Cheng-de & others. "A Taxonomical Study of *Phyllostachys,* China." *Acta Phytotaxonomica Sinica* Vol. 18, Nos. 1 and 2. 1980. Beijing.

Wen, Tai-hui. "A New Genus and Some New Species of *Bambusoideae* from China." *Journal of Bamboo Research* Vol. 1, No. 1. 1982. Hanzhou.

Wen, Tai-hui. "Studies on Bamboo Genus *Sinobambusa* from China and Other Species." (1) and (2). *Journal of Bamboo Research* Vol. 1, Nos. 1 and 2. 1982. Hanzhou.

Yi, Tong-pei. "New Taxa of *Bambusoideae* from Xizang (Tibet) China." *Journal of Bamboo Research* Vol. 2, No. 1. 1983. Hanzhou.

Yi, Tong-pei. "New Species of *Fargesia* Franchet and *Yushania* Keng f. from Tibet." *Journal of Bamboo Research* Vol. 2, No. 2. 1983. Hanzhou.

蕭灑臨風

癸亥仲秋為大鈞先生著作寫意自隆

雪海銀洲

癸亥年冬月 晓钢

3 KEY AND DESCRIPTIONS *of Important Genera and Species of Chinese Bamboos*

As bamboos do not flower for many years, classifying them by flower characteristics has little significance in practical horticulture and landscaping. In this chapter we will, therefore, base the keys and descriptions on the morphological characteristics of the vegetative organs.

MORPHOLOGICAL CHARACTERISTICS OF BAMBOOS

UNDERGROUND RHIZOME

According to the shape and construction of rhizomes, bamboos can be classified into four groups.

1. Monopodial Rhizomes with Scattered Culms (Stoloniferous group): Rhizomes slender with diageotropical growth, node prominent, roots grown on node, bud formed on the side of each node, which may grow into new rhizome or sprout as new shoot. Culms scattered above ground. For example, the genera *Phyllostachys* Siebold et Zuccarini and *Fargesia* Franchet belong to this group.

2. Sympodial Rhizomes (Clumped group): Rhizomes thick and shortened. The top buds emerge above ground into new shoots. Culms closely clumped. This group includes the genera *Bambusa* Schreber, *Sinocalamus* McClure, *Lingnania* McClure, *Schizostachyum* Nees, *Dendrocalamus* Nees, etc.

3. Sympodial Rhizomes with Scattered Culms (Intermediate group): Elongated rhizome necks travel underground before they emerge. Produce diffuse culms above ground. The genus *Sinarundinaria* Nakai belongs to this group.

4. Amphipodial Rhizomes (Combined Group): Rhizomes both monopodial and sympodial. Scattered culms interwoven with clumped culms. The genera *Pseudosasa* Makino and *Indocalamus* Nakai have such characteristics.

CULM

The culm is linked with the rhizome by a small, shortened part called the Culm Neck, without any roots attached. The Base of the Culm (Duckfoot) is composed of several to a dozen nodes with shortened thick internodes, has a strong root system around it. Culms round, erect, hollow inside, numerous nodes. From the upper nodes, branches develop. Close above the node is the nodal ridge, and under the node is a ring of sheath scar.

BRANCH

The branching habit of bamboos is also an important feature for classification.

1. One Branch Pattern: Solitary branch at node. The genera *Indocalamus* Nakai and *Sasamorpha* Nakai belong here.

2. Twin Branch Pattern: Two branches at node, one is dominant. The genus *Phyllostachys* Siebold et Zuccarini belongs here.

3. Triple Branch Pattern: Three branches at node, the central one is dominant. The genera *Sinobambusa* Makino and *Chimonobambusa* Makino belong here. But other genera, such as *Pseudosasa* Makino, which have two to four twigs on both side branches, making a clump of five to seven branches altogether at the upper nodes, also belong here.

4. Multi-Branch Pattern: Such genera as *Sinocalamus* McClure, *Bambusa* Schreber, *Lingnania* McClure, *Schizostachyum* Nees belong here.

SHEATH AND LEAF

Culm sheath papyraceous or leathery, deciduous or persistent, with or without auricles. Sheath blade undeveloped, without prominent midrib, deciduous or persistent. Ligule existent. The main part of a sheath, excluding the sheath blade is the sheath proper.

One leaf at node on the branchlet. Leaves arranged alternatively into two ranks. Leaf sheath with or without leaf auricles. Ligule membranaceous. Leaf blade with prominent midrib.

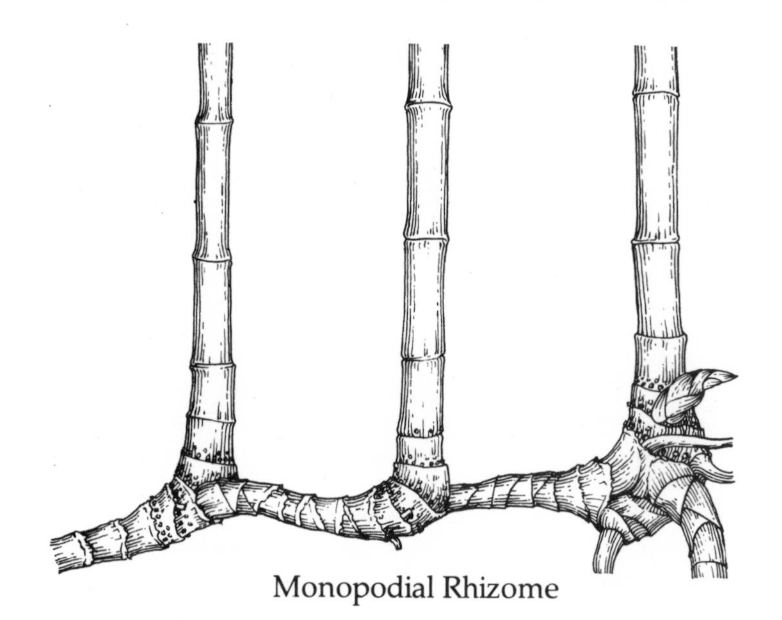

Monopodial Rhizome

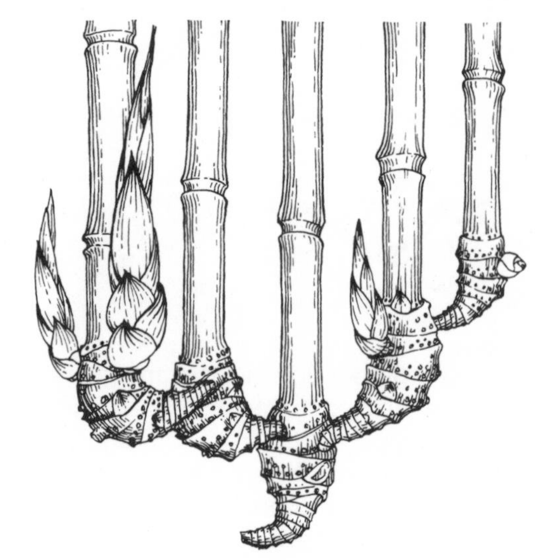

Sympodial Rhizome

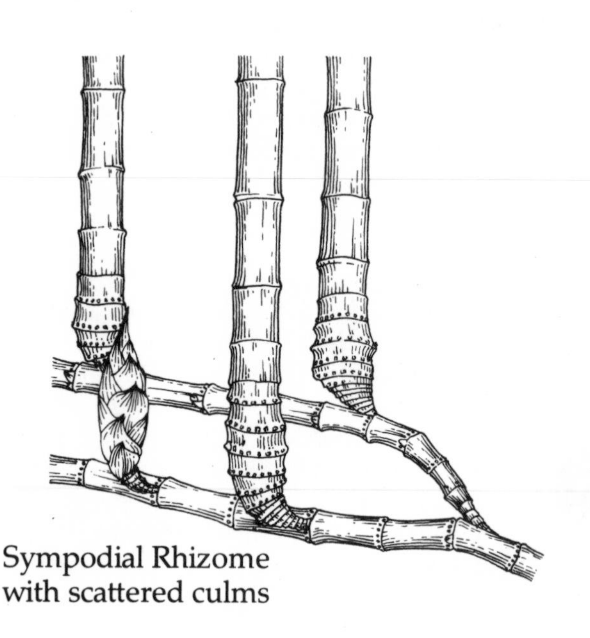

Sympodial Rhizome
with scattered culms

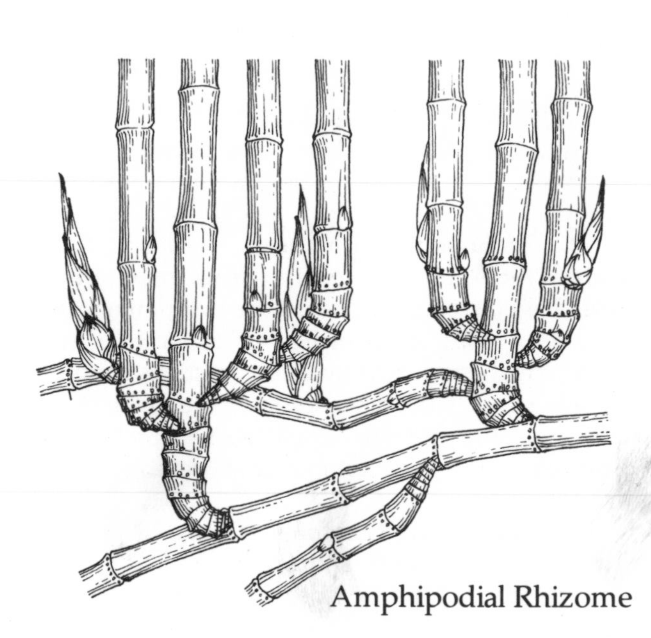

Amphipodial Rhizome

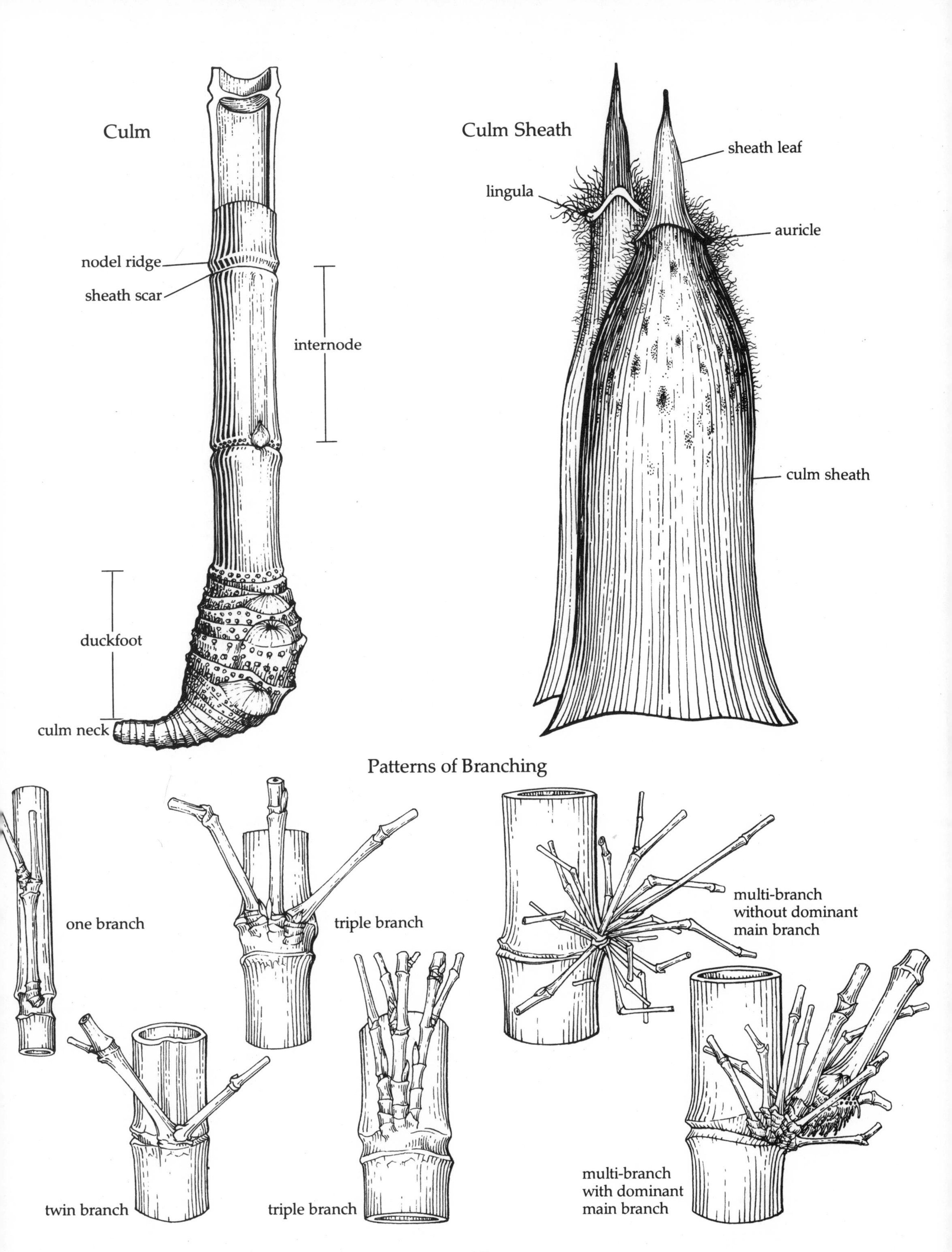
Culm
nodel ridge
sheath scar
internode
duckfoot
culm neck
Culm Sheath
sheath leaf
lingula
auricle
culm sheath
Patterns of Branching
one branch
triple branch
multi-branch
without dominant
main branch
twin branch
triple branch
multi-branch
with dominant
main branch

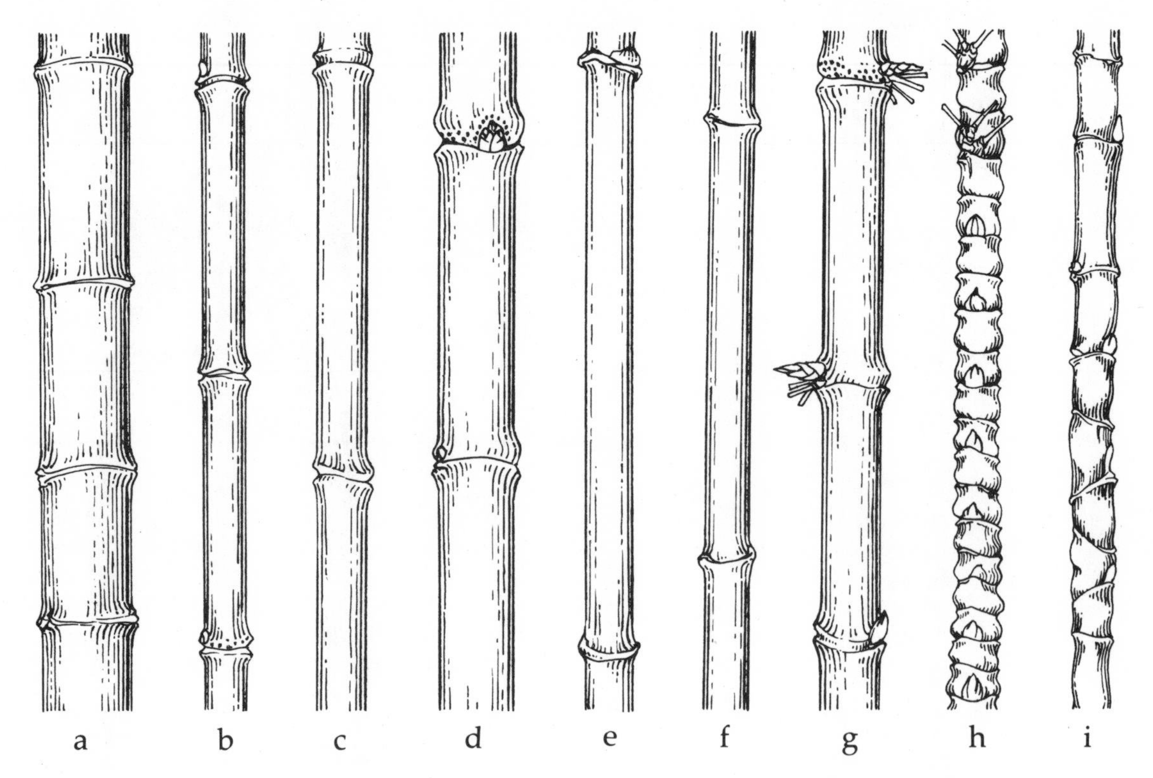

Nodes and Internodes

Forms, sizes and lengths in proportion. a. *Phyllostachys pubescens* Mazel ex H. de Lehaie. b *Phyllostachys glauca* McClure. c *Pseudosasa amabilis* (McClure) Keng f. d *Sinocalamus latiflorus* (Munro) McClure. e *Lingnania chungii* (McClure) McClure. f *Bambusa textilis* McClure. g *Bambusa pervariabilis* McClure. h *Bambusa ventricosa* McClure. i *Phyllostahcys aurea* Carr. ex Riv.

Key to the Genera of *Bambusoideae* Native to China

1a. Rhizomes monopodial or amphipodial. Culms diffuse or both diffuse and cespitose. 2
1b. Rhizomes sympodial (rhizome necks short or long, without buds and roots). 18
2a. Culm internodes cylindrical, rarely slightly compressed or grooved on the base of branching side. 3
2b. Culm internodes generally somewhat compressed or grooved on branching side, sometimes square. 14
3a. Primary branch single (rarely many) on nodes almost as thick as culm. Leaf blades large....... 4
3b. Primary branches 3 or many (rarely 1 to 2) on nodes. 7
4a. Dwarf bamboo. Culms about 1 m tall. 5
4b. Medium or tall bamboo. Culms 5 to 7 (9) m tall. Primary branch single on midculm nodes, fastigiate, almost parallel to the culm. Primary branches many on upper culm nodes. Branches swollen on base, aerial-root-like-verrucose. ...... I. *Ferrocalamus* Hsueh et Keng f.
5a. Culm sheaths longer than internodes (at least those on lower-culm). Sheath auricles and oral setae obsolescent or obsolete. .... II. *Sasamorpha* Nakai
5b. Culm sheaths generally shorter than internodes, with oral setae or with obvious sheath auricles. 6
6a. Nodal ridges not swollen. Primary branch single on nodes (rarely 2 to 3 on culm top nodes). Leaf blades lanceolate or long-lanceolate, the length over 4 to 5 times the breadth. .... III. *Indocalamus* Nakai
6b. Nodal ridges swollen. Primary branch 1 only on nodes. Leaf blade generally ovate-lanceolate, the length not over 4 times the breadth. ...... IV. *Sasa* Makino et Shibata
7a. Primary branches generally 3 on nodes. 8
7b. Primary branches many (rarely 1 to 3) on node. 11
8a. Culms diffuse. 9
8b. Culms diffuse or both diffuse and cespitose. Culm sheaths abscising early or late. 10
9a. Culm sheaths abscising early, glabrous. Primary branches 3 (rarely 5) on nodes. Leaves 2 to 5 (9) on twigs. Leaf blade lanceolate. .... V. *Brachystachyum* Keng
9b. Culm sheathes abscising late, densely pannose-pubescent. Primary branches 3 or more on midculm nodes, but generally 1 to 3 on upper-or lower-culm nodes. Leaves 4 to 10 on twigs. Leaf blades oblong-lanceolate or linear-lanceolate. .... VI. *Semiarundinaria* Makino
10a. Culms glabrous, not glaucous. Internodes slightly square only, solid or nearly solid near culm base. Primary branches usually 3 on nodes, or becoming many as secondary branches develop. Leaf blades lanceolate to narrow-lanceolate. .... VII. *Qiongzhuea* Hsueh et Yi
10b. Culms usually glaucous. Internodes near culm base cylindrical and fistulose. Primary branches usually 3 or many, rarely 1 to 2, on nodes. .... VIII. *Arundinaria* Michaux
11a. Culms diffuse and cespitose. 12
11b. Culms diffuse and cespitose, rarely diffuse only. Small or shrub-like bamboos. Culm sheaths abscising late or persistent. Nodal ridges prominently swollen. Sheath scars with sheath callus and fugacious-brown bristles, glabrate later. Primary branches 3 to 7 on midculm nodes. .... IX. *Pleioblastus* Nakai
12a. Small shrub-like bamboos. Primary branches 5 to 12 on nodes, without secondary branches. 13
12b. Tall or shrub-like bamboos. Culm sheaths late deciduous or persistent. Nodal ridges not swollen. Primary branches 1 to 3 on nodes, with secondary

branches. Leave 1 to many on twigs (secondary branches). .... X. *Pseudosasa* Makino

13a. Culm sheaths early deciduous. Nodal ridges prominently swollen. Primary branches generally (3) 5 to 7 on nodes, short, with 1 to 2 leaves on tip. .... XI. *Shibataea* Makino

13b. Culm sheaths persistent. Nodal ridges slightly swollen. Primary branches 7 to 12 on nodes, slender, spreading, usually with one leaf on tip. .... XII. *Gelidocalamus* Wen

14a. Culms diffuse. Branch nodes not swollen or slightly swollen. 15

14b. Culms diffuse or both diffuse and cespitose. Branch nodes prominently swollen. ... 17

15a. Primary branches generally 2 on nodes. Internodes grooved to the full length on the branching side. Culm sheaths early deciduous. .... XIII. *Phyllostachys* Siebold et Zuccarini

15b. Primary branches generally 3 on nodes. 16

16a. Culm sheaths persistent. Aerial roots (verrucose-spiny) on nodes of culm base. Several internodes near culm base somewhat square or cylindrical. Primary branches 3 at first, then increased to many later, on nodes. .... XIV. *Chimonobambusa* Makino

16b. Culm sheaths deciduous. No aerial roots (verrucose-spiny) on nodes of culm base. Internodes near culm base not square. Primary branches 3 on nodes (5 on upper-culm nodes). .... XV. *Acidosasa* C.D. Chu et C.S. Chao

17a. Culms diffuse and cespitose. Internodes long. Sheath callus on sheath scars as swollen as nodal ridge. Nodes depressed. Primary branches generally 3 (sometimes 5 to 7) on nodes, compressed, of equal size. .... XVI. *Sinobambusa* Makino ex Nakai

17b. Culms diffuse or both diffuse and cespitose. Nodal ridges prominently swollen, generally geniculate. Primary branches 3 (rarely 5) on nodes, sometimes 1 to 2 on lower culm nodes. The central branch thicker than the side branches. .... XVII. *Indosasa* McClure

18a. Rhizomes sympodial, with elongated necks. Culms pluricespitose and diffuse. 19

18b. Rhizomes sympodial, without elongated necks. Culms unicespitose. 20

19a. Elongated rhizome necks thick and solid, not very long. Primary branches 3 to many (rarely 1 to 2) on nodes. Culm sheaths generally coriaceous. Leaf blades generally small. .... XVIII. *Fargesia* Franchet

19b. Elongated rhizome necks slender, slightly fistulose, long-running. Primary branches 3 to 7 (rarely less or many) on nodes. Culm sheaths thin. Leaf blades very small. .... XIX. *Yushania* Keng f.

20a. The base of sheath blades almost as broad as the top of sheath proper. Sheath blades mostly upright; those reflexed usually bearing deformed spine-like twigs. Primary branches many (rarely a few) on nodes. .... XX. *Bambusa* Schreber

20b. The base of the sheath blades much narrower than or as broad as one-half of the top of sheath proper. Sheath blades usually reflexed. Twigs not deformed to spines. 21

21a. Internode surfaces and the abaxial surfaces of culm sheaths not siliceous. 22

21b. Internode surfaces and the abaxial surface of culm sheaths siliceous; or culms climbing in case of not siliceous. 26

22a. Culm internodes pruinose or none. 23

22b. Culm internodes sparsely pruinose only when young. 25

23a. Tall or shrub-like bamboos. Culms erect, rarely climbing. Culm sheaths deciduous. Sheath auricles none or obsolescent. 24

23b. Small shrub-like bamboos. Culms erect, climbing and drooping at top. Internodes not pruinose. Primary branches 2 to 3, to many on nodes. Culm sheaths late deciduous or persistent. Sheath auricles and leaf auricles with radiating long setae. . . . . XXI. *Ampelocalamus* Chen, Wen et Sheng

24a. Culm tops usually drooping. Internodes thickly pruinose. No aerial root thorns on nodes. Primary branches many on nodes. . . . . XXII. *Lingnania* McClure

24b. Culm tops upright. Internodes usually not pruinose, rarely gray-farinose. A girdle of aerial root thorns on nodes from midculm up. Primary branches 3 or many on nodes. Branch nodes prominently swollen. . . . . . XXIII *Chimonocalamus* Hsueh et Yi

25a. Giant bamboos. Culm tops usually drooping. Short aerial roots on culm base nodes. Primary branches many on nodes. Leaf blades generally large, oblong-lanceolate to narrow lanceolate. . . . . XXIV. *Dendrocalamus* Nees

25b. Tall bamboos. Culm tops arching or drooping. Generally no aerial roots on culm base nodes. Primary branches many on nodes. Leaf blades long and wide, broad-lanceolate to rectangular-lanceolate. . . . . . . XXV. *Sinocalamus* McClure

26a. Culms erect. Culm tops drooping or climbing. Primary branches many on nodes, of equal size. . . . 27

26b. Tall climbing bamboos. Internodes on lower culm zigzag. Primary branches many on nodes, the central one bigger than the others. . . . . XXVI. *Dinochloa* Buse

27a. Shrub-like bamboos. The part near top of sheath proper transversely protruded on the abaxial surface to a roundish-arcuate protuberance. No oral setae. Sheath blades upright. . . . . XXVII. *Leptocanna* Chia & Fung

27b. Tall or shrub-like bamboos. The part near top of sheath proper not protruded. Oral setae prominent. Sheath blades reflexed. . . . . XXVIII. *Schizostachyum* Nees

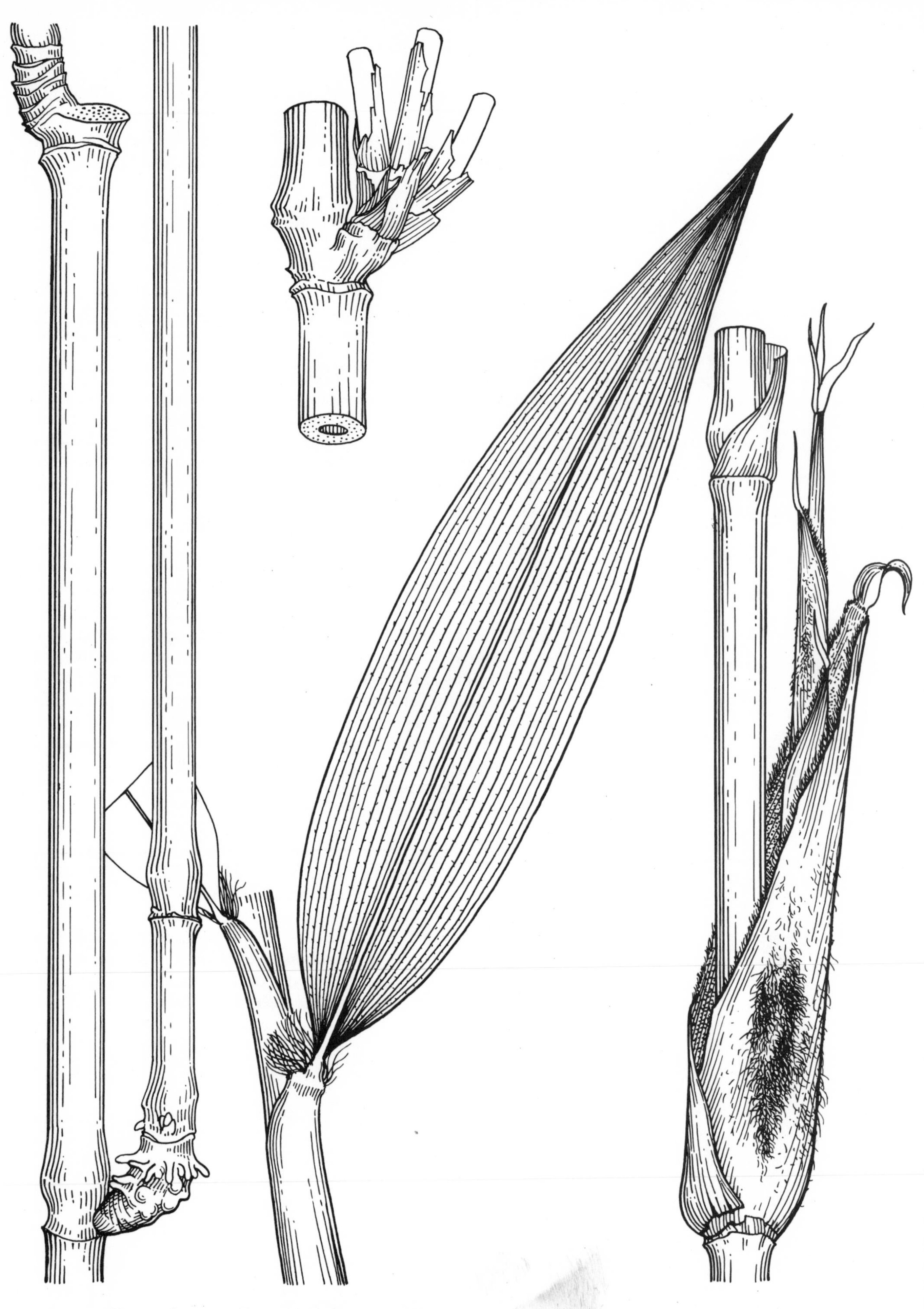

Ferrocalamus strictus Hsueh et Keng f.

I. *Ferrocalamus* Hsueh et Keng f., gen. nov.

Rhizomes monopodial. Culms diffuse. Primary branch 1 on midculm nodes, strong and erect, parallel to the culm; many on upper-culm nodes. Base of branches swollen, with aerial-root-like tubercules. The first internode of branches very short, 2 cm long only, internodes from second up greatly prolonged, sometimes reaching 65 cm long.

Ferrocalamus strictus Hsueh et Keng f., sp. nov.

Medium or small tree-like bamboo, rhizomes monopodial, slender. Culms 5 to 7 (9) m tall, tops erect, 2 to 3.5 (5) cm in diameter. Old culms green, but becoming pale yellowish-green when grown on mountain ridges. Culm wall thick, nearly solid at base. Internodes often 60 to 80 cm long, the longest reaching 120 cm. Intranodes 2 to 3 cm broad, lucid. Nodal ridges ribbed, prominently swollen on the opposite side of branching. Culm sheaths abscising late to persistent. Culm sheaths thick-coriaceous, except upper part being papery; crisp, densely blackish-brown setose and white-pilosulose, verrucose after setae are shed. The base of culm sheaths pale yellow-setose, retrorse-appressed to the node. Sheath ligules short, truncate at apex, rusty-pubescent on abaxial surface. Sheath auricles none. Oral setae prominent. Sheath blades upright, leaf-like-lanceolate, fugacious. Leaf blades large 30 to 55 cm long, 6 to 9 cm wide; long-acute at apex, cuneate at base; serrulate; grayish-green beneath. Midrib lucid, yellow, prominently elevated beneath, especially on lower part. Lateral veins 11 (10 to 12) pairs. Transverse veinlets in square pattern, elevated beneath.

Distributed at an altitude of 900 to 1200 m in Jinping Prefecture of Yunnan Province. Culm wall hard, used as arrow-shaft. Tender.

II. *Sasamorpha* Nakai

Shrub-like bamboos. Culms often less than 2 m tall, usually about 1 m. Rhizomes amphipodial. Culms erect, diffuse or pluricespitose. Internodes cylindrical, of uneven length on upper-culm; glabrous or pubescent, usually pruinose. Nodal ridges not swollen. Primary branch 1 on nodes, strong, nearly as thick as the culm. Branching on upper culms. Culm sheaths thick and hard, coriaceous, persistent, late deciduous, usually pubescent and pruinose, longer than internodes and tightly appressed to the culm; but loosened by branches and consequently appressed tightly to the branch. Sheath auricles and oral setae obsolescent or obscure. Leaves 2 to 3 (5) on top of twigs. Leaf blades lanceolate, glabrous or sparsely hairy beneath. Leaf sheaths coriaceous, without leaf auricles and oral setae. Grown under trees. Dense clumps forming a thick ground cover.

KEY to Spp. of *Sasamorpha* Native to China

1a. Culm sheaths densely persistent, long-tuberculated-based-hairy. Branch sheaths long-radiating-hairy. .... (1) *S. qingyuanensis*

1b. Culm sheaths glabrous or fugacious-sparse-verruciform-pilose, or persistent verruciform-pilose on top of culm sheaths only. 2

  2a. Culm sheaths and top of branch sheaths comparatively thin and rugose. Leaf sheaths densely white-appressed-villous, glabrous under nodes. .... (2) *S. hubeiensis*

  2b. Culm sheaths and top of branch sheaths not rugose. Sparsely short-hairy under nodes. .... (3) *S. sinica*

Sasamorpha sinica (Keng) Koidz.

1. ***Sasamorpha qingyuanensis*** C. H. Hu, sp. nov.

Culms 1 to 1.5 m tall, 4 to 6 mm in diameter; cylindrical. Internodes on upper culms usually more or less shortened between branching nodes; pruinose, more prominently under nodes, forming a girdled-thick-pruinose-zone. Culm sheaths persistent, densely brown- or white-verruciform-villous, appressed or nearly appressed. The base of sheath proper with a girdle of densely brown setae. Leaves 3 on twigs. Leaf blades long-oblong to long-ovate, 18 to 28 cm long, 4.7 to 6 cm wide; long-acuminate at apex, somewhat caudate; gradually narrowed at base to stalks; green above, pale green beneath; glabrous; entire or indistinct-minute-serrulate on one side of rim only. Lateral veins 10 to 13 pairs. Transverse veinlets distinct.

Distributed at an altitude of 1400 m in Qingyuan Prefecture of Zhejiang Province. Hardy to −8°C.

2. ***Sasamorpha hubeiensis*** C. H. Hu, sp. nov.

Culms 0.5 to 1 m tall, 0.3 to 0.5 cm in diameter; yellow or pale yellow; lucid; with thick pruinose zone under nodes. Internodes usually obviously shortened and of unequal length on upper culms between branching nodes. Culm sheaths persistent, longer than internodes, somewhat lucid, thin-coriaceous. Primary branches semi-upright; branch sheath fugacious verruciform-pilose; densely white-pubescent at base, mingled brown-pilose. Leaves 3 on top of twigs. Leaf sheaths sparsely pruinose, densely appressed-upwards-villous. Leaf blades lanceolate (the lowest elliptic), gradually narrow at base to stalks (the lowest truncate at base); semi-coriaceous, glabrous. Lateral veins 7 to 9 pairs, tertiary veins 5 to 7 between every two lateral veins. Transverse veinlets distinct. Reticular veins square in form.

Distributed at an altitude of 1300 m in Mount Jiugongshan in Tonghsan Prefecture of Hubei Province. Hardy to −15°C.

3. ***Sasamorpha sinica*** (Keng) Koidz.

[*Sasamorpha sinica* f. *glabra* C. H. Hu]

Culms 1.5 m tall, 4 mm in diameter. Culm sheaths persistent, 6 to 10 cm long, pale purple, glabrous or setulose on upper part, ciliate. Oral setae absent. Primary branch 1 on nodes, with 1 to 2 leaves. Leaf sheaths glabrous. Leaf blades oblong to lanceolate, 2 to 4 cm wide; grayish beneath, pubescent at base. Lateral veins 6 to 8 pairs. Transverse veinlets exserted.

Distributed at an altitude of about 1000 m in Zhejiang and Anhui Provinces. Hardy to −10°C.

Leaf blades glabrous and smooth.

III. *Indocalamus* Nakai

Arbuscular or small arbuscular bamboos, rhizomes monopodial or amphipodial. Culms erect, diffuse or pluricespitose. Internodes cylindrical, nodes broad, not or insignificantly prominent. Culm sheaths persistent, usually shorter than internodes. Primary branch 1 on nodes, seldom 2 to 3 on culm-top nodes, fastigiate. The diameter of branch nearly equal to that of culm. Leaf blade large, quite broad, lanceolate or long lanceolate, lateral veins many, transverse veinlets distinct.

KEY to the Spp. of *Indocalamus* Native to China

1a. Culms dwarf, under 1 m tall. Leaf blades 0.9 to 2 cm wide. Lateral veins 4 to 7 pairs. 2
1b. Culms ovgr 1 m tall. Leaf blades 1.5 to 5 cm or more wide. Lateral veins over 6 pairs. 3
  2a. Culms 30 cm tall. Nodes slightly hairy. No pruinose zones under nodes. . . . . (1) *I. pedalia*
  2b. Culms 90 cm tall. Pruinose zones under nodes. . . . . (2) *I. varius*
    3a. The width of leaf blades not over 4.5 cm. Lateral veins 5 to 10 pairs. 4
    3b. The width of leaf blades over 5 cm. Lateral veins 6 to 12 (18) pairs. 6
      4a. Culm internodes hollow. 5
      4b. Culms solid or nearly solid, 3 m tall. Leaf blades narrow-lorate-lanceolate, 20 to 30 cm long, 3 to 4 cm wide, lateral veins about 10 pairs. . . . . (3) *I. solidus*

5a. Culms 1 to 1.5 m tall. Leaf blades lanceolate, 14 to 23 cm long, 2.5 to 4 cm wide. Lateral veins 5 to 9 pairs. . . . . (4) *I. victorialis*

5b. Culms 2 m tall. Leaf blades oblong-lanceolate, 16 to 32 cm long, 2 to 4.5 cm wide. Lateral veins 8 to 10 pairs. . . . . (5) *I. pseudosinicus*

6a. The abaxial surfaces of culm sheaths and leaf sheaths usually glabrous. Culm internodes nearly solid. Culm sheaths much longer than internodes. A distinct line of pannose-pubescence along one side of midrib on the back surface of leaf blades. Lateral veins 15 to 18 pairs. . . . . (6) *I. tessellatus*

6b. The abaxial surfaces of culm sheaths and leaf sheaths brown-setulose adnate, at least when young. Culm sheaths shorter than internodes. 7

7a. Culm sheath auricles very prominent. 8

7b. Culm sheath auricles and leaf auricles not prominent. 10

8a. Leaf auricles prominent. 9

8b. Leaf auricles very small or obsolescent. Culms 1.8 m tall. Leaf blades 10 to 28 cm long, 4.5 to 7.5 cm wide. Lateral veins 9 to 13 pairs. . . . . . (7) *I. hunanensis*

9a. Culms about 1 m tall. Leaf blades 10 to 30 cm long, 1.5 to 6.5 cm wide. Lateral veins 6 to 12 pairs. . . . . (8) *I. longiauritus*

9b. Culms about 1.5 m tall. Leaf blades 18 to 40 cm long, 3.5 to 6.5 cm wide. Lateral veins about 10 pairs. . . . . (9) *I. emeiensis*

10a. Culms about 1 m tall. 11

10b. Culms 2 m tall. Culm sheaths coriaceous. Leaf blades dark green and slightly hairy (when young) above, whitish-green and puberulent beneath. . . . . (10) *I. migoi*

11a. Culm sheaths thin and soft, not adnate, erect-setose on the abaxial surfaces. Upper margin of sheath ligules glabrous, nearly truncate. Leaf blades green and sparsely hairy above, whitish-green and pubescent evenly beneath. . . . . (11) *I. lacunosus*

11b. Culm sheaths fragile, adnate, appressed-setose on the abaxial surfaces. Upper margin of sheath ligules arcunate, fimbricate. Leaf blades green and glabrous above, whitish-gray and pilosulose beneath. . . . . (12) *I. latifolius*

1. *Indocalamus pedalia* (Keng) Keng f.

Diageotropical rhizome internodes 5 to 20 cm long, 2 mm in diameter. Roots 3 on nodes.

Culms 30 cm tall, cespitose. Lower internodes 1 to 5 cm long, 1 to 1.5 mm in diameter. Leaves 2 to 3 on tops. Usually one suberect primary branch on base nodes. Leaf ligules ciliate. Leaf blades coriaceous or papery, lanceolate, 6.5 to 15 cm long, 9 to 17 mm wide; acuminate at apex, rounded at base; glabrous, or puberulent beneath when young. The upper rim of leaf blade serrulate, the lower rim nearly smooth. Lateral veins 4 to 6 pairs, transverse veinlets distinct. Leaf stalks 2 to 4 mm long.

Distributed in Sichuan Province. Usually grown in crevices of rocks. Hardy to −5°C.

2. *Indocalamus varius* Keng f.

Culms 90 cm tall, 3 mm in diameter. Internodes 14 cm long. Primary branches 1 to 2 on nodes. Leaves 2 to 3 on top of twigs. Leaf sheaths lineate, with transverse veinlets. Oral setae several, yellow. Leaf ligules hard, truncate. Leaf blades lanceolate, 5 to 11 cm long, 10 to 20 mm wide, acuminate at apex, narrowed or rounded at base, puberulent beneath, serrulate, stalked. Lateral veins 4 to 7 pairs, transverse veinlets exserted.

Distributed in hilly regions of Zhejiang Province. Hardy to −7°C.

3. *Indocalamus solidus* C. D. Chu et C. S. Chao, sp. nov.

Culms 3 m tall, 8 to 10 mm in diameter, arching at top. Young culms green. Old culms blackish-green, glabrous, slightly pruinose only under nodes. Midculm internodes 40 to 50 cm long. Culm sheaths adnate to culms, glabrous except purplish-brown-ciliate on rims. Sheath auricles prominent, amplexical, oral setae 1.5 cm long. Sheath blades triangular-lanceolate, not clasping, fugacious. Primary branch 1 on nodes, compressed. Leaf sheaths pruinose on upper part, glabrous, purplish-brown-ciliate on rims. Leaf auricles prominent, oral setae remain. Leaf ligules very short, ciliolate on upper rim. Leaf blade thick, acuminate at apex, truncate or nearly so at base, nearly sessile, whitish-green beneath, glabrous.

Distributed in calcareous hilly areas of Yangsu Prefecture, Guangxi Province. Tender.

4. *Indocalamus victorialis* Keng f.

Culms 1 to 1.5 m tall; cylindrical. Internodes smooth, slender. Nodal ridges slightly swollen. Culm sheath coriaceous or chartaceous, villous on lower part, retrorse-brown-villous at base, ciliate on margins. Sheath ligules rounded, truncate at top. Sheath blades slender, glabrous, fugacious. Primary branch 1 (seldom 3 or 4) on nodes, adnate to culm. Leaf sheaths glabrous except ciliate on upper margins, liratus on abaxial surfaces. Leaf ligules truncate, ciliate. Leaves 1 to 4 on twigs, papery, glabrous; acuminate to apiculate; narrowed on base to stalk; green above, grayish-green beneath.

Distributed in Fujian and Sichuan Provinces. Tolerates light frost.

5. *Indocalamus pseudosinicus* McClure

Culms 2 m tall. Primary branch single on nodes. Branch internodes slender, cylindrical, glabrous except densely appressed-pubescent under nodes. Branch nodes swollen above sheath scars, glabrous. Sheath scars slightly swollen. Branch sheaths somewhat persistent. The apex of leaf blades narrowed abruptly, acuminate to awned. The base of leaf blades narrowed abruptly. Leaf blades scabrous beneath, especially near apex. Transverse veinlets distinct only on the under surface. Leaf sheaths glabrous, brown-ciliate, with veins slightly elevated, convex at apex with both corners round-shoulder-like. Leaf auricles obscure, oral setae 6 to 9. Leaf ligules very short.

Distributed in Hainan Island of Guangdong Province. Tender.

6. *Indocalamus tessellatus* (Munro) Keng f.

Culms 1 to 2 m tall, 0.5 to 1 cm in diameter. Internodes 2.5 to 5 cm long. Culm sheaths longer than internodes. Sheath ligules arcuate, a few setae on both sides. Sheath blades slender, varied in size. Primary branch 1 (seldom 2) on nodes. Leaf blades long-lanceolate, the largest over 45 cm long, over 10 cm wide; green and glossy above, grayish-green and sparsely gray-puberulent beneath, midrib elevated. Transverse veinlets distinct.

Distributed in hilly areas of 1000 m high in the provinces of Yangtze Valley. Hardy to −12°C.

7. *Indocalamus hunanensis* B. M. Yang, sp. nov.

Culms 1.8 m tall. Culm internodes 20 to 26.7 cm long, yellowish-green, longitudinal-lineate, strewn with yellowish-brown or brown bristles based on tubercles, and white pubescence, especially dense under nodes of young culms. Tubercles remaining on old culms after shedding bristles. Nodal ridges not distinct. Sheath scars distinct with corky sheath callus. Culm sheaths persistent, hard and fragile, reddish-brown, as long as ⅓ to ⅔ of internodes. On the abaxial surface of sheath, white-puberulent; on its lower part strewn with yellowish-brown or brown tubercle-based bristles; brown ciliate on rims. Sheath auricles strongly developed, long-falciform, clasping or spirally twisted; purple, turning blackish-purple when dry; sheath ligules very short, slightly arcuate, whole or minutely lobed on rim, glabrous or ciliolate, extremely short. Sheath blades erect or those on lower nodes reflexed, linear-lanceolate.

Primary branch 1 on nodes except a few near culm base. The base of branches adnate. Leaves 3 to 5 on twigs, on lower branch nodes, leaf sheaths usually without blades or with small ovate-lanceolate blades. Leaf blades oblong-lanceolate, dark-green above, grayish-green beneath, scabrous on rims. Leaf apex acuminate. Leaf base rounded-or-broad-cuneate, narrowed to stalk.

Distributed in Hunan Province. Tolerates slight frost.

8. *Indocalamus longiauritus* Hand.-Mazz.

Culms about 1 m tall, 5 mm in diameter. Internodes 5 to 20 cm long, glabrous. Primary branches 1 to 3 on nodes. On the abaxial surface of culm sheaths strewn with appressed brown bristles. Sheath ligules truncate. Oral setae exserted. Sheath auricles distinct, half-moon like, fimbricate-ciliate. Sheath blades triangular. Leaf sheaths glabrous, or strewn with small appressed brown bristles when young. Leaves 1 to 3 on twigs. Leaf auricles distinct, half-moon like, fimbriate-ciliate. Lateral veins 6 to 12 pairs, transverse veinlets exserted.

Distributed among woods or shrubberies in eastern or southwestern China. Hardy to −7°C.

9. *Indocalamus emeiensis* C. D. Chu et C. S. Chao, sp. nov.

Culms about 1.5 m tall; hollow. Young culms white-appressed hispidulous. Internodes 30 cm long, reddish-brown barbed on upper parts. Culm sheaths persistent, shorter than ½ of internodes, brown, densely brown-barbed, regularly brown-ciliate. Sheath auricles developed, luniform. Oral setae radiating. Sheath blades triangular-lanceolate. The base of sheath blade about half the width of

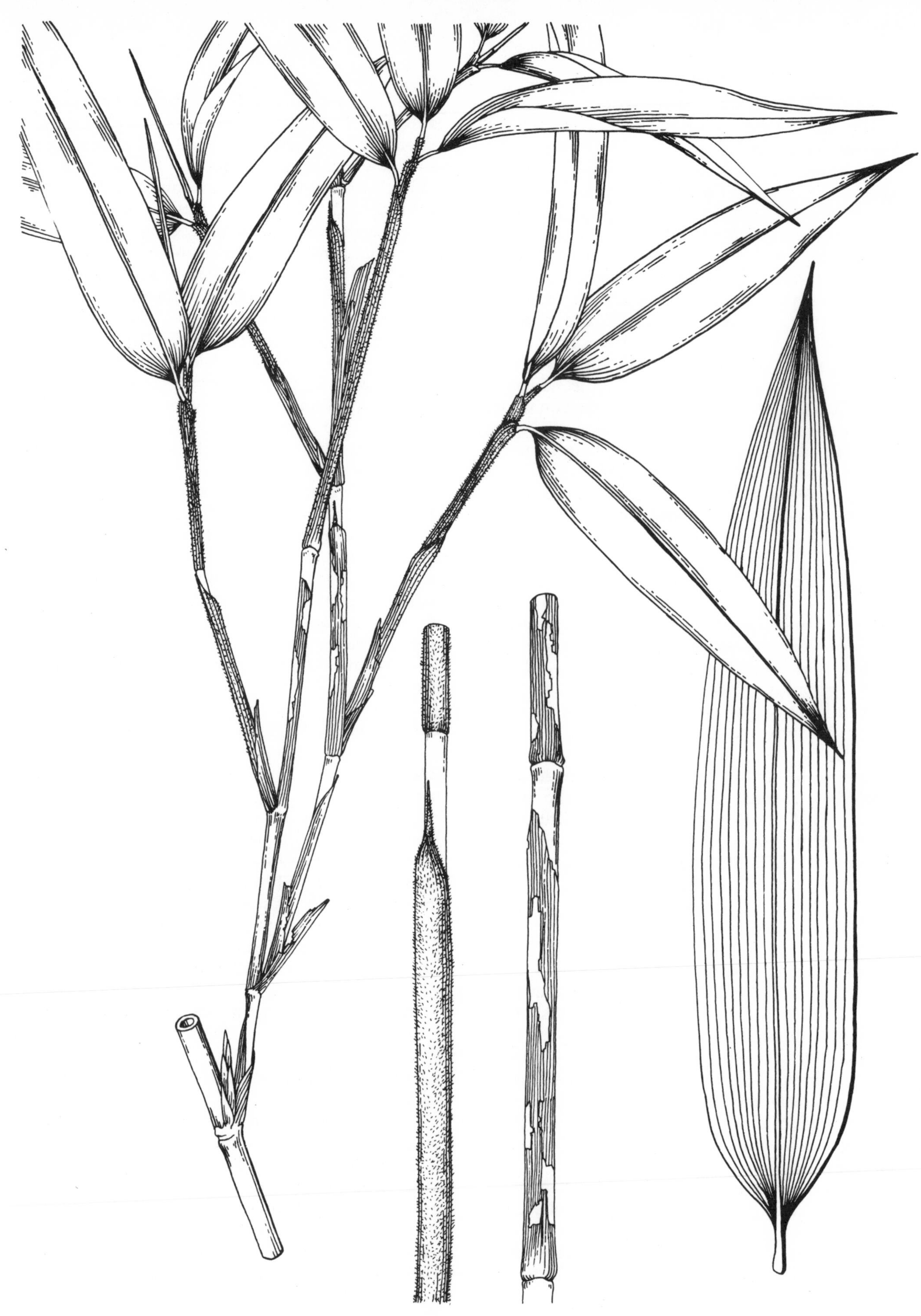

Indocalamus migoi (Nakai) Keng f.

sheath ligule. Sheath ligules about 1 mm long, thick-ciliate at apex. Primary branch single on nodes. Leaves 4 to 10 on twigs. Leaf sheaths white-pubescent and also reddish-brown-hispidulous, rims ciliate. Leaf auricles developed, luniform. Oral setae radiating. Leaf ligules short, 3 cm long fimbriate bristles on the rim at apex. Leaf blades broad-lorate-lanceolate, leaf apex acuminate, leaf base cuneate, narrowed to stalk. Both upper part and base of leaf blades usually somewhat asymmetrical. The whole blade glabrous, whitish-green beneath, serrulate on both sides. Lateral veins about 10 pairs.

Distributed in Mount Emei of Sichuan Province at an altitude of 1200 m. Hardy to −8°C.

10. *Indocalamus migoi* (Nakai) Keng f.
[*Sasamorpha migoi* Nakai]

Culms 2 m tall, 1 to 1.5 cm in diameter. Nodal ridges not swollen. Culm sheaths coriaceous, purplish-brown, setulose and also white-lanulose on abaxial surfaces; becoming glabrescent when old. Oral setae about 5 mm long. Sheath ligules 1 to 2 mm long, indumentum on abaxial surfaces. Sheath auricles very small or absent. Leaf sheaths usually yellowish-brown-setulose, glabrate when old. Leaf blades lanceolate to oblong-lanceolate, 25 to 50 cm long, 4 to 6 cm wide; slightly hairy above, glabrate later and becoming glossy and dark green; grayish-green and puberulent beneath. Transverse veinlets exserted.

Distributed under woods in hilly areas of Jiangxi and Zhejiang Provinces. Hardy to −7°C.

11. *Indocalamus lacunosus* Wen, sp. nov.

Culms 1 m tall, 8 mm in diameter. Nodes not swollen. Culms not farinose, glabrous except some fine hairs on nodes of young culms. Leaves 5 to 7 on twigs. Leaf sheaths 7 to 8 cm long, glabrous. Leaf blades elliptic to ovate, 15 to 30 cm long, 3 to 5 cm wide; rounded at base, asymmetrical; leaf apex cuspidate, procurrent to caudate, serrulate on rims. Lateral veins 8 to 10 pairs, transverse veinlets absent.

Distributed in Chong-an Prefecture of Fujian Province. Tender.

12. *Indocalamus latifolius* (Keng) McClure
[*Arundinaria latifolia* Keng]

Culms about 1 m tall, densely clumped or scattered, 5 mm in diameter. Internodes 5 to 20 cm long, slightly hairy. Branches 1 to 3 on nodes. Culm sheaths hard, brownish-black setulose on abaxial surfaces. Sheath ligules truncate, 0.5 to 1 mm long, sheath auricles none, oral setae 1 to 3 mm long. Leaves 1 to 3 on twigs. Leaf blades long-lanceolate, 10 to 40 cm long, 1.5 to 8 cm wide, bright green and glabrous above, grayish-white and slightly hairy beneath. Lateral veins 6 to 12 pairs, transverse veinlets exserted.

Distributed in Mount Qinling of Shaanxi Province, central and eastern China, at an altitude below 1000 m. Hardy to −15°C; tolerates partial shade.

IV. *Sasa* Makino & Shibata

Shrub-like or sub-shrub-like bamboos. Rhizomes monopodial or amphipodial. Culms slender, cylindrical. Primary branch 1 on nodes. Culm sheath persistent or deciduous, coriaceous, glabrous or hairy. Leaf blades comparatively broad, oblong-lanceolate, coriaceous or thick-papery, glabrous. Transverse veinlets distinct.

KEY to the Spp. of *Sasa* Native to China

1a. Culm sheath auricles, leaf sheath auricles and oral setae absent. 2
1b. Culm sheath auricles, leaf sheath auricles and oral setae remain. 5
  2a. Culms about 1 m tall. 3
  2b. Culms 2 to 3 m tall. Internodes longer, densely setose. Leaf blades large, elliptic-lanceolate to lorate-lanceolate, 25 to 35 cm long, 3 to 8 cm wide. . . . . (1) *S. bashanensis*
    3a. Leaf blades smaller. 4

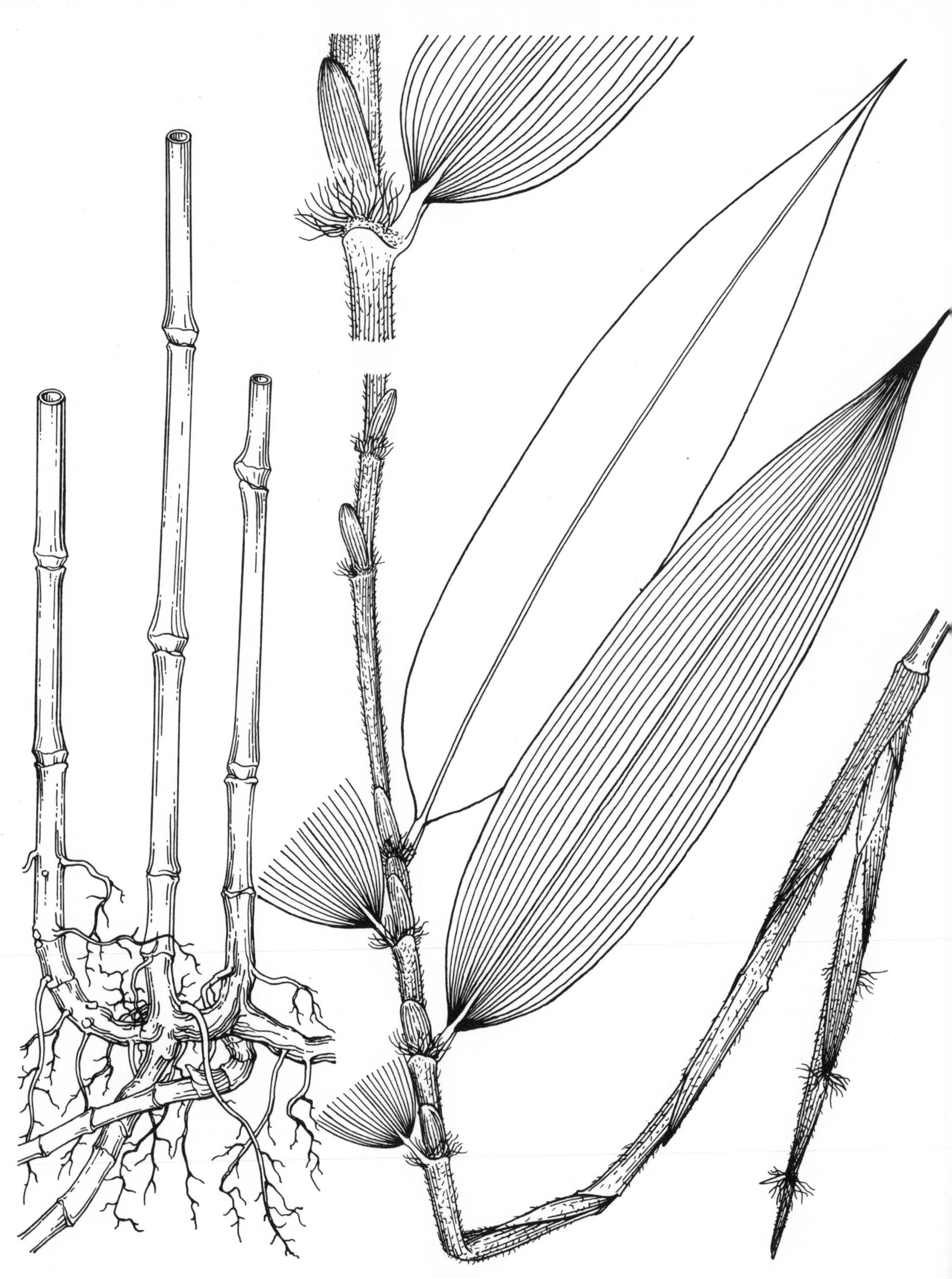

Sasa guangxiensis C.D. Chu et C.S. Chao

3b. Leaf blades large. Leaves 6 on twigs. Leaf blades ovate-lanceolate to broad-lorate-lanceolate, 18 to 32 cm long, 6 to 7 cm wide, entire, slightly undulate; lateral veins indistinct beneath. . . . . (2) *S. hainanensis*

4a. Nodal ridges prominently swollen. Primary branches 15 to 30 cm long, slant-fastigiate or slightly spreading, with secondary branches. Leaves 3 to 15 on twigs. Leaf blades lanceolate, 6 to 25 cm long, 15 to 33 mm wide. . . . . (3) *S. longiligulata*

4b. Nodal ridges slightly swollen. Primary branches short, upright, without secondary branches. Leaves 3 to 5 on twigs. Leaf blades broad-lanceolate, 6 to 15.8 cm long, 18 to 47 mm wide. . . . . (4) *S. nubigena*

5a. Culms 1 m tall. Nodal ridges strongly swollen, geniculate. Culm sheaths longer than or as long as internodes, pubescent on abaxial surface. Leaf ligules especially prominent. . . . . (5) *S. guangxiensis*

5b. Culms 2 m tall. Nodal ridges slightly swollen, not geniculate. Culm sheaths as long as 1/2 to 1/3 of internodes, densely tomentose on abaxial surface. Leaf ligules extremely short. . . . . (6) *S. tomentosa*

1. *Sasa bashanensis* C. D. Chu et C. S. Chao, sp. nov.

Rhizomes monopodial or amphipodial. Culms 2 to 3 m tall, 1 to 1.5 cm in diameter. Midculm internodes 38 to 42 cm long. From mid-internode up densely fugaciously porrect-setose, pitted after shedding of setae; pruinose. Nodal ridges prominently swollen. Culm sheaths densely brown-porrect-setose. Sheath ligules 2 to 4 mm long, nearly truncate at apex. Sheath blades short, narrow-lanceolate. Leaf ligules prominent, 4 to 7 mm long, arcuate at apex, entire or slightly undulate on rim, not ciliate. Leaf blades serrulate. Lateral veins 11 to 13 pairs.

Distributed at an altitude of 600 to 1200 m on calcareous mountainous areas in Shaanxi Province. Hardy to −15°C.

2. *Sasa hainanensis* C. D. Chu et C. S. Chao, sp. nov.

Culms 1 to 1.5 m tall; 0.5 to 1 cm in diameter; small and dwarf. Leaf sheaths glabrous. Leaf ligules 3 mm long, truncate at apex, entire. Leaf blades acuminate at apex, rounded and somewhat oblique at base; entire, slightly undulate; pale green or powdery green beneath; glabrous; sparsely pruinose.

Distributed at an altitude of 1000 m in Hainan Island of Guangdong Province. Tender.

3. *Sasa longiligulata* McClure

Culms semi-erect, slightly arching at top, 1.5 m tall, 1 cm in diameter. Sheath scars densely glochidiate. Intranodes grayish-white-retrorse-setulose. Culm sheaths loosely encircled internodes. Leaf blades of thick-papery texture, acuminate and subulate at apex, contracted and asymmetrical at base; lucid above.

Distributed at an altitude of 900 to 1000 m on calcareous soil with abundant moisture in Guangdong Province. Tender.

4. *Sasa nubigena* Keng f.

Culms 60 cm tall, 2 to 3 mm in diameter. Internodes smooth and glabrous; cylindrical or slightly grooved at base on budded side. Sheath scars distinct. Culm sheaths tightly encircle internodes. The lowest leaf on every twig with leaf sheath ciliate on outer rim. Leaf blade apiculate, green above, yellowish-green beneath; glabrous.

Distributed at an altitude of 1800 m on Mount Jinfushan in Sichuan Province. Hardy to −10°C.

5. *Sasa guangxiensis* C. D. Chu et C. S. Chao, sp. nov.

Culms erect, 1 m tall, 5 mm in diameter; green when young; pale-yellow-pubescent, especially denser under nodes, glabrous later. Culm sheaths deciduous. Sheath auricles lunate, oral setae prominent, 5 to 10 mm long, radiating. Sheath ligules obvious, 5 mm long, lacerate at apex. Sheath blades lanceolate, erect, fugacious. Leaves 3 to 8 on twigs. Leaf sheaths densely pubescent, glabrous later. Leaf auricles distinct, lunate; setae radiating, 6 to 10 mm long. Leaf blades elliptic-lanceolate to lanceolate, 13 to 26 cm long, 2 to 4.5 cm wide; long-acuminate at apex, broad-cuneate at base; glabrous; grayish-white beneath. Lateral veins 4 to 7 pairs. Transverse veinlets distinct.

Distributed at an altitude of 500 to 1500 m in Guangxi Province. Tender.

6. *Sasa tomentosa* C. D. Chu et C. S. Chao, sp. nov.

Culms 2 m tall, 5 mm in diameter, fistulose. Internodes 15 to 22 cm long, cylindrical, glabrous. Nodal ridges strongly swollen, but not geniculate. Culm sheaths short, densely pale-yellow-verrucose-tomentose. Sheath auricles falciform, 5 mm long; setae radiating, 8 to 10 mm long. Sheath ligules short, slightly hairy. Sheath blades lanceolate, 1.5 to 3.5 cm long, 2 to 5 mm wide, upright or spreading. Leaves 2 to 3 on twigs. Leaf sheaths tomentose. Leaf auricles prominent; setae radiating, 10 to 13 mm long. Leaf blades elliptic-lanceolate to lanceolate, 18 to 20 cm long, 3 to 4.2 cm wide; acuminate at apex, broad-cuneate at base; glabrous grayish-white beneath. Lateral veins 8 to 9 pairs. Transverse veinlets distinct.

Distributed in Guangxi Province at an altitude of 1400 m. Tender.

V. *Brachystachyum* Keng

Small or shrub-like bamboo. Rhizomes monopodial. Culms diffuse, cylindrical or slightly compressed on branching side. Nodal ridges swollen. Culm sheaths abscising early.

Brachystachyum densiflorum (Rendle) Keng

Culms 2 m tall, 1 cm in diameter. Young culms pubescent, glabrous when old. Pruinose under sheath scars, becoming blackish-dirty later. Culm sheaths yellowish-green, not maculate or both white and purple striate, hirsute, purple-ciliate on rim. Sheath auricles prominent, purple or green. Sheath blades green. Leaf sheaths hard, with transverse veins. Oral setae several. Primary branches 3 (rarely 5) on every culm node. Leaves 2 to 5 on twigs. Leaf blades lanceolate, 5 to 18 cm long, 10 to 25 mm wide; dark green and glabrous above, grayish-brown and pilosulose beneath. Lateral veins 4 to 8 pairs. Transverse veinlets distinct.

Distributed in hilly areas around Lake Taihu. Hardy to −12°C.

Brachystachyum densiflorum var. ***villosum*** S. L. Chen et C. Y. Yao, var. nov.

Differing from the sp. by a mass of yellowish-brown hairs on the base of culm sheaths.

VI. *Semiarundinaria* Makino

Tall or shrub-like bamboos. Rhizomes monopodial. Culms erect. Internodes cylindrical, slightly compressed on branching side. Culm sheaths deciduous. Sheath blades linear to lanceolate. Primary branches 3 to 5 (usually 3) on midculm nodes, 1 to 2 (usually 1) on upper-culm and lower-culm nodes. Leaf blades oblong-lanceolate to linear-lanceolate. Leaf auricles and leaf ligules not properly developed. Leaf auricles setose on rim.

KEY to the Spp. of *Semiarundinaria* Native to China

1a. Culm sheaths usually glabrous on abaxial surface, except girdled-villous on base united to the sheath node. Young culms purplish-brown, pruinose zone under nodes. .... (1) *S. shapoensis*
1b. Culm sheaths fugacious-setose on abaxial surface. Young culms not pruinose. 2
  2a. Culms 2 m tall, young culms pale purple. Culm sheaths grayish-white-setose on abaxial surface, hispid on base. .... (2) *S. gracilipes*
  2b. Culms 5 m tall, young culms green. Culm sheaths sparsely white to pale-yellow-setose on abaxial surface, glabrescent on base. .... (3) *S. lubrica*

1. *Semiarundinaria shapoensis* McClure

Culms 2 m tall, 1 cm in diameter. Internodes glabrous, purplish-brown when young, pruinose under nodes. Nodes swollen, glabrous. Primary branches of nearly even thickness on each node, or the central one stronger and longer than the others. Leaves 5 to 10 on twigs. Leaf blades linear-lanceolate, 5 to 13.5 cm long, 6 to 15 mm wide; slightly rugose; long-acuminate and scabrous subulate at apex, broad-cuneate to nearly rounded at base; glabrous above, scabrid beneath; setulose on rims. Transverse veinlets exserted.

Distributed in Hainan Island of Guangdong Province. Tender.

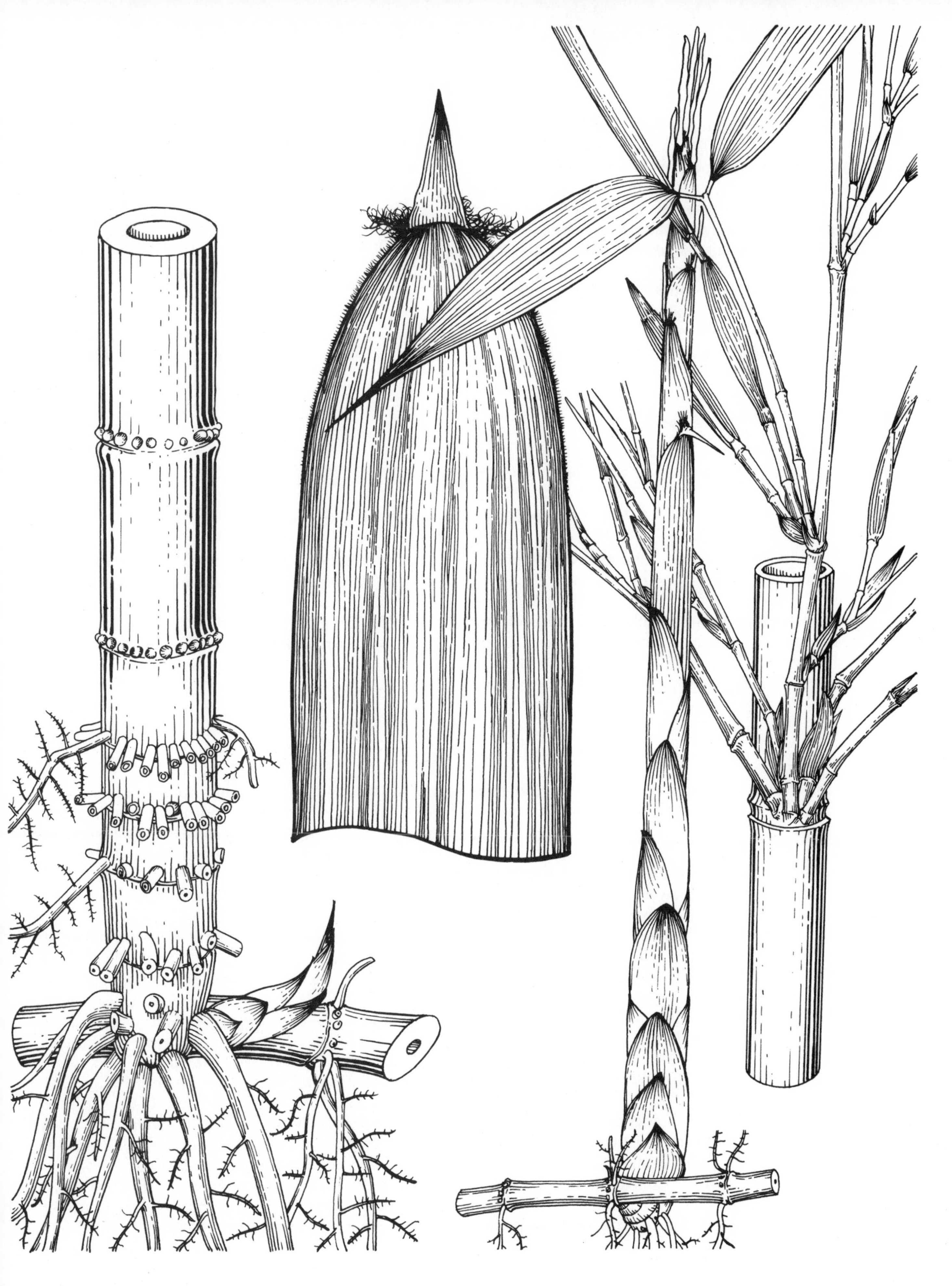

Brachystachyum densiflorum (Rendle) Keng

2. *Semiarundinaria gracilipes* McClure

Culms 2 m tall, 1 cm in diameter at base. Internodes pale purple when young. Nodes swollen, glabrous. Primary branches usually 3 on nodes near culm base, less on midculm or upperculm nodes, usually spreading. Leaf blades linear-lanceolate to oblong-lanceolate, 10 to 15 cm long, 9 to 20 mm wide; acuminate and subulate at apex, cuneate or nearly rounded at base; glabrous, scabrous on rims. Lateral veins distinct, 4 to 5 pairs. Transverse veinlets exserted.

Distributed in Hainan Island of Guangdong Province. Tender.

3. *Semiarundinaria lubrica* Wen, sp. nov.

Culms 5 m tall, 2 cm in diameter. Internodes 30 cm long. Primary branches 3 on nodes, of nearly even thickness, compressed. Leaves 3 to 4 on twigs. Leaf sheaths white-pubescent, with longitudinal veins. Leaf auricles purple, with radiating setae. Leaf blades lanceolate, 10 to 15 cm long, 15 to 22 mm wide; obtuse at base, acute and caudate at apex, smooth or scabrous on both sides of rim, glabrous or scabrous beneath only. Lateral veins 6 pairs. Transverse veinlets distinct, in square pattern.

Distributed in Zhejiang Province. Hardy to −5°C.

VII. *Qiongzhuea* Hsueh et Yi, gen. nov.

Shrub-like bamboos. Rhizomes amphipodial. Culms erect. Internodes cylindrical or slightly square only on several basal internodes; compressed on branching side, glabrous, not glaucous, solid or nearly solid on lower internodes. Culm wall rather thick. Nodal ridges not swollen to slightly swollen or strongly swollen to becoming rounded-ribbed. Culm sheaths of thick papery texture. Primary branches usually 3 on nodes, becoming many-branched as secondary branches develop, fastigiate to spreading. Leaf blades lanceolate to narrow-lanceolate. Transverse veinlets distinct.

KEY to the Spp. of *Qiongzhuea* Native to China

1a. Internodes on culm base cylindrical. Nodal ridges strongly swollen, and rounded-ribbed, much thicker than internodes. Leaf blades narrow-lanceolate, 5 to 14 cm long, 6 to 12 mm wide, lateral veins 2 to 4 pairs. . . . . (1) *Q. tumidinoda*

1b. The several lowest internodes near culm base cylindrical or sometimes slightly square. Nodal ridges on unbranching nodes not swollen or slightly swollen. Leaf blades lanceolate. 2

2a. Culm sheaths glabrous or sometimes sparsely setulose on abaxial surface, longitudinal veins indistinct. Primary branches usually 3 on nodes. 3

2b. Culm sheaths sparsely brown-setulose on abaxial surface, scabrid when young, longitudinal veins distinct. 4

3a. Leaves 2 to 3 on twigs. Leaf blades (5) 8 to 12 cm long, (8) 13 to 20 mm wide, pilosulose on abaxial surface. Lateral veins 4 to 5 pairs. . . . . (2) *Q. communis*

3b. Leaves usually 1 to 2 on twigs. Leaf blades 7 to 13 cm long, 8 to 17 mm wide, glabrous on abaxial surface. Lateral veins (3) 4 pairs. . . . . (3) *Q. rigidula*

4a. Primary branches usually 2 to 3 on nodes. Leaves 1 (2) on twigs. Leaf blades 7.5 to 17 cm long, (10) 13 to 16 mm wide, pilosulose beneath. Lateral veins 4 to 5 pairs. . . . . (4) *Q. opienensis*

4b. Primary branches usually 3 on nodes. Leaves 2 to 3 (4) on twigs. Leaf blades glabrous. 5

5a. Culm sheaths abscising late. Young culms pubescent (denser under nodes). Branches cylindrical or slightly triangular near base, pubescent under nodes. Leaf blades papery 10 to 15 cm long, 10 to 16 mm wide. Lateral veins (3) 4 to 5 (6) pairs. . . . . . (5) *Q. puberulla*

5b. Culm sheaths persistent. Internodes glabrous. Branches cylindrical, glabrous under nodes. Leaf blades thick-papery, 15 to 23 cm long, 16 to 20 mm wide, lateral veins 5 to 7 pairs. . . . . (6) *Q. luzhiensis*

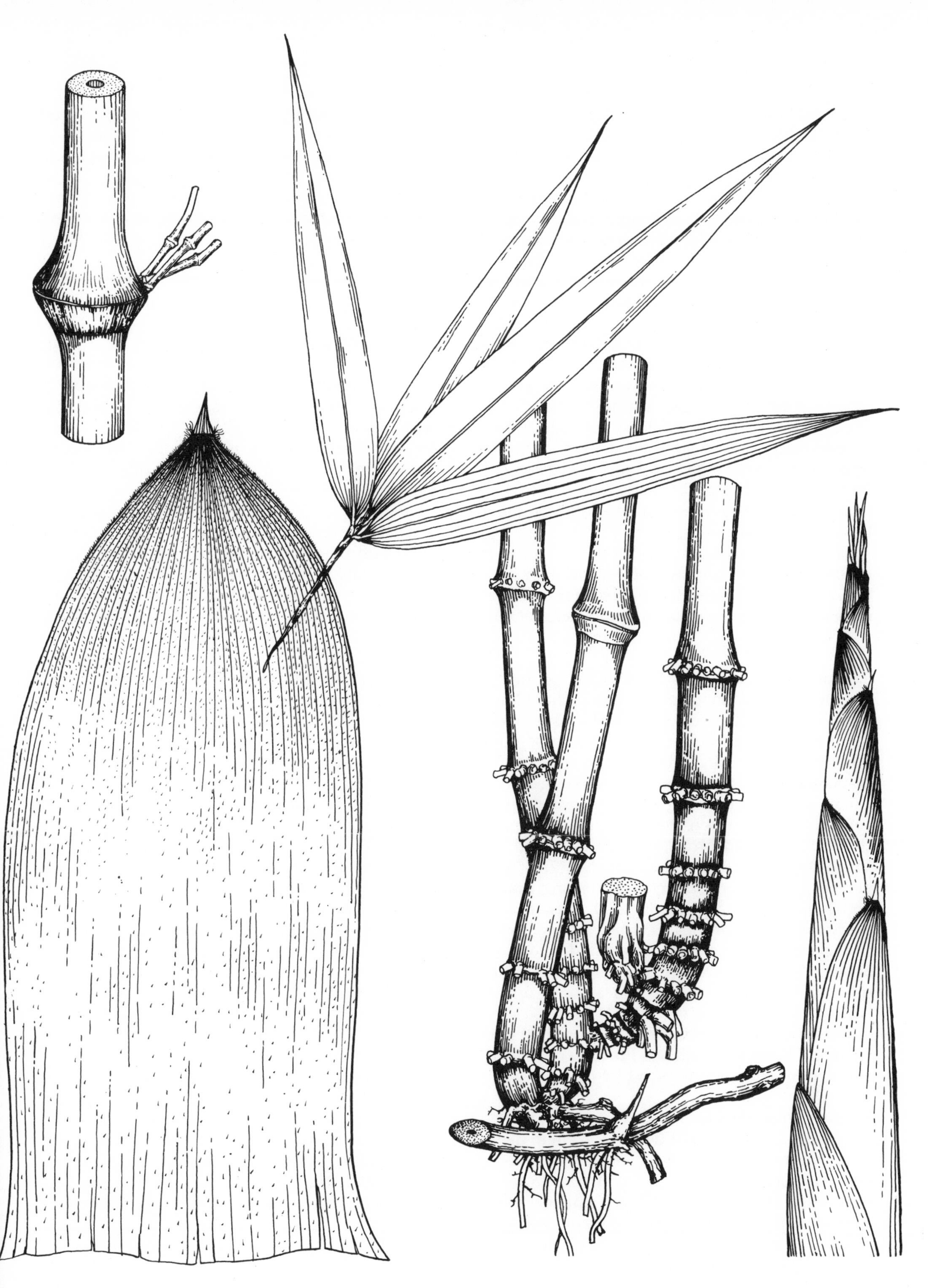

Qiongzhuea tumidinoda Hsueh et Yi

1. *Qiongzhuea tumidinoda* Hsueh et Yi, sp. nov.

Rhizomes amphipodial. Culms 2.5 to 6 m tall, 1 to 3 cm in diameter. Internodes on lower-culm-unbranched part usually single narrow-grooved on the same side; those on upper-culm compressed on branching side with 2 longitudinal ribs and 3 grooves; 15 to 25 cm long (except those near culm base 10 to 15 cm long only). Sheath scars with sheath callus, brown-setose when young, glabrous later. Culm sheaths early deciduous, stramineous, thick papery, about as long as ½ of internodes, longitudinal veins many and distinct on abaxial surface, brown-verruciform-setose. Upper rims of culm sheaths densely pale-brown-ciliate. Oral setae brown, 2 to 3 mm long. Sheath auricles none. Sheath blades 5 to 17 mm long, subulate or awl-shaped-lanceolate, upright. Primary branches 3 on nodes, with lateral branches 1 to 4. Twigs slender, 20 cm long, 2 mm in diameter. Leaves 3 to 4 on twigs, leaf blades green above, grayish-green beneath, glabrous.

Distributed at an altitude of 1500 to 2100 m in northeastern Yunnan and southwestern Sichuan provinces. Hardy as the native habitats often frozen in winter months with a minimum temperature −15°C.

2. *Qiongzhuea communis* Hsueh et Yi, sp. nov.

Rhizomes monopodial or amphipodial. Culms 3 to 7 m tall, 1 to 3 cm in diameter. Internodes slightly square or cylindrical near culm base, compressed, 2-longitudinal-ribbed and 3-grooved on branching side. Sheath scars distinct, glabrous. Culm sheath blackish-green, becoming pale-yellowish-brown when shedding. Sheath auricles absent. Sheath blades triangular or awl-shaped, 5 to 11 cm long, glabrous. Primary branches (2) 3 (5) on nodes, spreading or horizontally spreading. Branch nodes swollen. Twigs nearly solid, clustered on every node on branches. Leaf blades dark green and glabrous above, pale green and pilosulose beneath, densely serrulate and scabrous on one side, while sparsely serrate and smooth on the other.

Distributed at an altitude of 1600 to 1980 m in Sichuan, Hubei, and Guizhou Provinces. Tolerates a minimum temperature of −15°C.

3. *Qiongzhuea rigidula* Hsueh et Yi, sp. nov.

Culms 2 to 4 (6) tall, 1.5 to 2.5 (3) cm in diameter. Internodes (10) 15 to 18 (24) cm long, glabrous. Culm wall 4 to 10 cm thick; 25 to 31 internodes on whole culm. Sheath scars glabrous. Nodal ridges on branching nodes slightly swollen. Culm sheaths early deciduous, thick-papery to leathery, densely yellow-brown ciliate. Sheath auricles obscure, oral setae absent. Sheath blades triangular or awl-shaped, glabrous, fugacious. Leaf blades serrulate on one side only.

Distributed at an altitude of 1300 to 1700 m in Sichuan Province. Tolerates a minimum temperature of −15°C.

4. *Qiongzhuea opienensis* Hsueh et Yi, sp. nov.

Culms 2 to 7 m tall, 1 to 5.5 cm in diameter. Internodes 18 to 20 (25) cm long, but those near base 6 to 10 cm long only; green, turning to yellowish-green when old, glabrous, not glaucous; 1-to-3-longitudinal-ribbed and 2–5-grooved on branching side. Usually branching from the 12th node up. 20 to 35 internodes on whole culms. Sheath scars glabrous; nodal ridges swollen on branching nodes only. Culm sheaths abscising early, thick-papery to leathery, yellowish-brown ciliate from the middle up; smooth and lucid on adaxial surface. Sheath auricles none. Sheath ligules purplish-brown, 1 mm long. Branch nodes prominently swollen. Twigs, slender, 1 to 5 on every node of branches. Leaf blades serrulate and scabrous on one side, while scabrid or nearly smooth on the other.

Distributed at an altitude of 1600 to 1900 m in Sichuan Province. Tolerates a minimum temperature of −15°C.

5. *Qiongzhuea puberulla* Hsueh et Yi, sp. nov.

Rhizomes amphipodial. Culms 4 to 5 m tall, 1.5 to 2.5 cm in diameter. Internodes (8) 15 to 17 (20) cm long, cylindrical or slightly square, 2-longitudinal-ribbed and 3-grooved on branching side; green or purplish-green, not glaucous. 25 to 32 internodes on whole culms. Young culms puberulent, denser under nodes. Sheath scars distinct, brown, brown-setulose. Nodal ridges slightly swollen, or swollen only on branching nodes, glabrous, smooth and lucid. Culm sheath abscising late, leathery, yellowish-brown, brown-setose on abaxial surface, densely brown- or yellowish-brown-setose on rims, longitudinal veins distinct. Sheath auricles none, oral setae 2 to 3, grayish, slightly curving, 1 to 4 mm long. Sheath ligules 1 mm long. Sheath blades upright, triangular, 2 to 13 mm long, 1 to 2 mm wide, purple when young. Primary branches 3 (7) on nodes, 20 to 65 cm long, with 3 to 11 nodes; usually purple. Branch nodes swollen, pubescent under nodes. Leaves (2) 3 (4) on twigs, papery. Leaf blades green above, grayish-green beneath, serrulate and scabrid on rim.

Distributed on yellow soil of calcareous origin

at an altitude of 1600 m in Liuzhi Special Region of Guizhou Prefecture. Tolerates a minimum temperature of −15°C.

6. *Qiongzhuea luzhiensis* Hsueh et Yi, sp. nov.

Rhizomes amphipodial. Culms 2.5 to 5 m tall, 1 to 2 cm in diameter, with 20 to 26 internodes. Internodes (10) 14 to 18 (20) cm long, cylindrical or slightly square; compressed, 2-longitudinal-ribbed and 3-grooved on branching side, (rarely with one groove only); green, glabrous, and not glaucous. Sheath scars swollen, yellowish-brown-setulose when young; nodal ridges swollen, glabrous, lucid. Culm sheaths persistent, leathery, reddish-brown or yellowish-brown; sparsely brown-setose on abaxial surface, with dense brown setae on rims. Sheath auricles none, oral setae 3 to 5, 2 to 5 mm long, fugacious. Sheath ligules yellowish-brown, 1 mm long. Sheath blades upright, triangular or linear-lanceolate, 2 to 7 mm long, purplish-brown or yellowish-brown. Primary branches 3 (5) on nodes, spreading, 20 to 60 cm long, with 8 to 10 nodes, swollen on branching nodes. Leaves 2 to 3 (4) on twigs. Leaf blades sparsely serrulate, and scabrid on rims.

Distributed at an altitude of 1700 to 1900 m in Liuzhi Special Region of Guizhou Province. Tolerates a minimum temperature of −15°C.

VIII. *Arundinaria* Michaux

Tall or shrub-like bamboos. Rhizomes monopodial or amphipodial. Internodes cylindrical, fistulose. Primary branches 3 to many (rarely 1 to 2) on nodes, the main branches dominant. Culm sheaths abscising late or early.

KEY to the Spp. of *Arundinaria* Native to China

1a. Culm sheaths hairy. 2
1b. Culm sheaths glabrous. 6
2a. Culm sheaths brown-setose. 3
2b. Culm sheaths yellowish-brown, densely chestnut-brown-glochidiate on abaxial surface, setose on base. Sheath ligules rounded, 5 mm long. Sheath blades slender, hard, and erect, rarely reflexed. . . . . (1) *A. amabilis*
3a. Sheath auricles and setae obscure. 4
3b. Sheath auricles and setae prominent. Culm sheaths brownish-yellow or greenish-yellow, margins purplish. Sheath ligules slightly arcuate or nearly truncate at apex, 2 to 4 mm long. Sheath blades lanceolate, green, or somewhat purplish. . . . . (2) *A. dushanensis*
4a. Leaf sheaths glabrous. 5
4b. Leaf sheaths brown- or pale-brown-setose when young. Culm sheaths not maculate. Sheath ligules slightly elevated, 2 mm long. Sheath blades shorter, lanceolate. . . . (3) *A. fargesii*
5a. Culm sheaths reddish-brown, becoming grayish-brown when dry. Sheath ligules short, 1 mm long. Sheath blades narrow-triangular to triangular-lanceolate, erect, slightly rugose. . . . . (4) *A. spongiosa*
5b. Culm sheaths green, becoming yellow when dry; purplish-brown maculate on abaxial surface. Sheath ligules 2 to 4 mm long. Sheath blades linear to linear-lanceolate, erect or spreading. . . . . (5) *A. maculosa*
6a. Primary branches many on nodes. Leaves 2 to 4 on twigs. Leaf blades 3 to 7.5 cm long, 4 to 14 mm wide. . . . . (6) *A. fangiana*
6b. Primary branches 1 to 2 on nodes. Leaves 2 to 3 on twigs. Leaf blades 6 cm long, 10 mm wide. . . . . (7) *A. flexuosa*

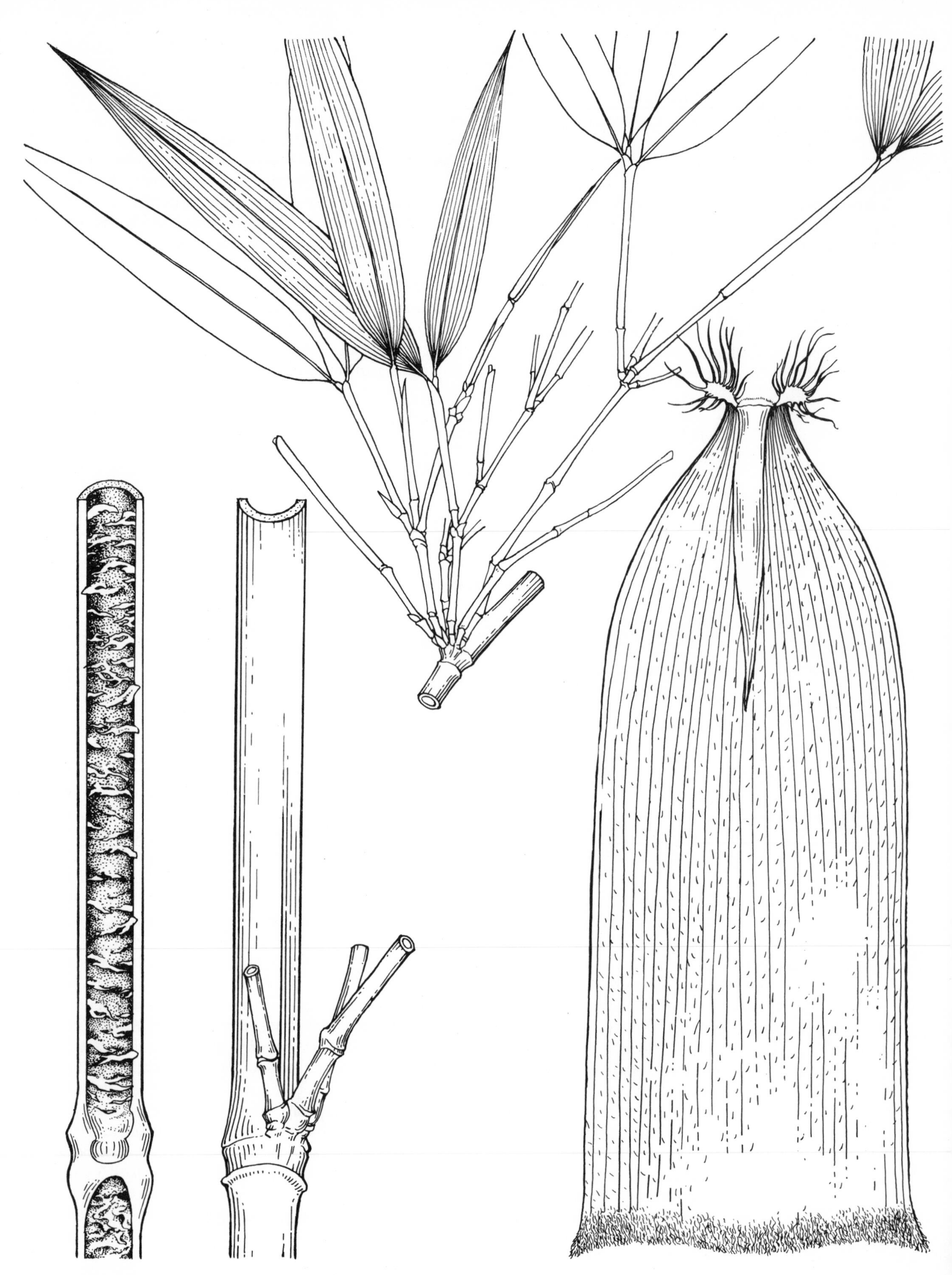

Arundinaria dushanensis C.D. Chu et J.Q. Zhang

1. *Arundinaria amabilis* McClure

Rhizomes monopodial. Culms diffuse, 6.5 m tall, rarely reaching 13 m, 5.7 cm in diameter, glabrous, grayish waxy on surfaces, distinctly striate, maculate when mature. Branching nodes slightly swollen. Internodes straight, cylindrical; longer on midculm and shorter on both ends. Primary branches 3 (sometimes single) on nodes, compressed, fastigiate, glabrous. Leaf blades linear-lanceolate, 18 to 35 cm long, 1.8 to 3.5 cm wide; acuminate at apex, gradually narrowed at base to stalks; pilosulose on base beneath. Transverse veinlets exserted.

Distributed in Guangdong and Guangxi Provinces. Tender.

2. *Arundinaria dushanensis* C .D. Chu et J. Q. Zhang, sp. nov.

Culms 10 m tall, 2.5 cm in diameter, green and glabrous when young. Culm wall 4 to 6 mm thick, sponge-like incrassate. Internodes 25 to 40 cm long, cylindrical, slightly grooved near branching nodes. Nodal ridges slightly swollen. Primary branches 3 (5 on upper-culm) on nodes. Leaves 2 to 3 on twigs. Leaf auricles and setae none. Leaf ligules protruding. Leaf blades lorate-lanceolate, 10 to 18 cm long, 1 to 2 cm wide, acuminate at apex, glabrous, powdery-green beneath.

Distributed in Dushan Prefecture of Guizhou Province. Tender.

3. *Arundinaria fargesii* E. G. Camus

Rhizomes monopodial or amphipodial. Culms diffuse or both diffuse and pluricespitose; 10 m tall, 4 to 5 cm in diameter. Culm wall thick, lumen small, sponge-like incrassate. Internodes on midculm 40 to 60 cm long. Nodal ridges slightly swollen. Sheath scars covered with girdle of brown hair. Primary branches many on nodes, main branches prominent. Branch sheaths persistent. Leaves 4 to 6 on twigs. Leaf auricles and setae obsolete. Leaf ligules protruding, elevated. Leaf blades lorate, lorate-lanceolate to long-ovate-lanceolate, vary greatly in size, 10 to 20 (30) cm long, 1 to 2.5 (5) cm wide; serrulate, sparsely hairy or glabrescent along veins beneath. Lateral veins 5 to 8 pairs.

Distributed in Mount Qingling in Shaanxi Province. Hardy to −15°C.

4. *Arundinaria spongiosa* C. D. Chu et C. S. Chao, sp. nov.

Culms erect, 10 m tall, 4 to 6 cm in diameter, green when young, glabrous, sparsely farinose under nodes; yellowish-green when old. Culm pith sponge-like. Internodes 20 to 40 cm long, cylindrical, slightly grooved near branching side. Nodes slightly swollen. Primary branches 3 on nodes. Leaves 3 to 5 on twigs. Leaf auricles and setae obsolete. Leaf ligules prominent, 2 to 2.5 mm long. Leaf blades lanceolate to linear-lanceolate, 9 to 17 cm long, 1 to 2 cm wide; long-acuminate at apex, glabrous. Lateral veins 4 to 5 pairs.

Distributed in Guangxi Province. Tender.

5. *Arundinaria maculosa* C. D. Chu et C. S. Chao, sp. nov.

Culms 10 m tall, 3.5 to 5.5 cm in diameter; green, slightly farinose and glabrous when young; yellowish-green or yellow when old. Culm pith ring-like incrassate. Internodes 30 to 40 cm long, slightly grooved near branching side. Nodes slightly swollen. Primary branches short, 3 on nodes. Leaves 2 to 5 on twigs. Leaf auricles and oral setae obscure. Leaf ligules prominently protruding. Leaf blades linear or linear-lanceolate, 5 to 13 cm long; 8 to 15 mm wide; long-acuminate at apex; glabrous. Lateral veins 3 to 4 pairs.

Distributed at an altitude of 500 to 1000 m on Mount Daqingshan and Mount Shiwandashan of Guangxi Province. Tender.

6. *Arundinaria fangiana* A. Camus

[*Sinarundinaria fangiana* (A. Camus) Keng ex Keng f.]

Small shrub-like bamboo. Culms 0.5 to 1 m tall, 4 to 7 mm in diameter, glabrous. Primary branches clustered as bundled on nodes. Leaves 2 to 4 on twigs. Culm sheaths glabrous except sometimes ciliate on rims; often shorter than internodes; oral setae obsolete. Leaf sheaths glabrous, with oral setae and leaf auricles. Leaf ligules 0.5 mm long, truncate at apex. Leaf blades acuminate-apiculate at apex, rounded and contracted at base to stalks; glabrous; scabrous on rims. Lateral veins 3 to 5 pairs. Transverse veinlets exserted.

Distributed at an altitude of 2500 to 3800 m in Sichuan, Shaanxi, Gansu, and Ningxia Provinces. Hardy to −20°C.

7. *Arundinaria flexuosa* Hance

Small shrub-like bamboo. Culm base appressed with imbricate culm sheaths. Culm sheaths grayish-yellow, glabrous, obtuse at apex. Sheath blades leaf-like. Primary branches 1 to 2 on nodes. Leaves 2 to 3 on branch tips. Leaf blades oblong-lanceolate, mucronate at apex, glabrous; smooth and lucid above, paler in color beneath; scabrous-serrulate. Transverse veinlets prominent.

Distributed in Guangdong Province. Tender.

IX. *Pleioblastus* Nakai

Small or shrub-like bamboos. Rhizomes generally amphipodial, a few monopodial. Culms diffuse or pluricespitose; erect; cylindrical or slightly compressed on branching side. Young culms green and pruinose. Old culms yellowish-green. Primary branches 3 to 7 on nodes. Nodal ridges prominently swollen, more elevated than sheath scars. Culm sheaths late deciduous or persistent. Sheath callus remains. Sheath blades subulate-lanceolate. Leaves 3 to 5 (8) on twigs. Leaf blades lanceolate, moderate in size. Leaf sheaths persistent, with oral setae. Lateral veins 5 to 7 pairs. Transverse veinlets distinct.

KEY to the Spp. of *Pleioblastus* Native to China

1a. Culms solid. Nodes prominently swollen, blackish-farinose under nodes. Culm sheaths pale green, with fugacious white bristles. .... (1) *P. solidus*

1b. Culms not solid or nearly solid. 2

2a. Culms sheaths pale brown, brown maculate, oily-lucid. .... (2) *P. maculatus*

2b. Culm sheaths generally green, not maculate or sparsely maculate; not lucid or not oily lucid. 3

3a. Sheath auricles prominent, with prominent setae. 4

3b. Sheath auricles and setae none. Sheath blades lanceolate. 10

4a. Sheath auricles falciform. 5

4b. Sheath auricles spoonform or elliptic. 9

5a. Culm sheaths setose on abaxial surface. 6

5b. Culm sheaths glabrous on abaxial surface, except brown-setose on base. 8

6a. Leaf ligules very short, membranous, 0.3 mm long. Culm sheaths dense-fugacious pruinose and purple-setulose on abaxial surface. Leaf auricles with reddish-purple setae on rim. Sheath blades lanceolate, spreading or reflexed. .... (3) *P. yixingensis*

6b. Leaf ligules ovate, 1 to 2 mm long. 7

7a. Culm sheaths hard, fugacious-setose; lanose on base. Sheath auricles with 8 mm long setae. Sheath blades triangular, erect, rugose. .... (4) *P. rugatus*

7b. Culm sheaths setose and pruinose; densely long-setose on rim of base. Oral setae upright, arranged in half-moon-like pattern. Sheath blades narrow-lorate, erect or reflexed. .... (5) *P. hsienchuensis*

8a. Culm sheaths persistent. Culms 3 to 4 m tall. Internodes 29 to 42 cm long. .... (6) *P. longifimbriatus*

8b. Culm sheaths abscising late. Culms 1.7 to 3 m tall. Internodes 20 to 28 cm long. .... (7) *P. juxianensis*

9a. Culm sheaths not maculate or sparsely maculate, glabrous, not pruinose; densely dark-yellowish-brown hirsute. Sheath ligules pale yellowish, concave at apex, glabrous. Sheath blades slender-lanceolate, drooping, with yellow margins. .... (8) *P. kwangsiensis*

9b. Culm sheaths oily-lucid, glabrous on abaxial surface, densely golden-velvety at base. Sheath ligules truncate or slightly arcuate at apex, glabrous or sometimes pubescent. Sheath blades lorate to subulate, reflexed. .... (9) *P. oleosus*

10a. Culm sheaths hairy. 11

10b. Culm sheaths glabrous. Sheath ligules 3 mm long. Sheath blades drooping. Leaf ligules about 3 mm long. ... (10) *P. altiligulatus*

11a. Culm sheaths purple-setose. Sheath ligules retuse or arcuate at apex. Sheath blades spreading or reflexed. Leaf ligules protruding, 5 mm long. .... (11) *P. intermedius*

11b. Culm sheaths fugacious, purplish-red-setulose. Sheath ligules truncate at apex. Sheath blades reflexed-drooping. Leaf ligules truncate at apex, 0.5 to 2 mm long. .... (12) *P. amarus*

1. *Pleioblastus solidus* S. Y.Chen, sp. nov.

Rhizomes slender. Culms erect, 4 to 5 m tall, 1.5 to 2 cm in diameter; greenish-yellow; hairy and pruinose when young. Internodes 24 to 33 cm long. Nodes swollen, blackish-farinose under nodes. Primary branches 5 on nodes, one main branch thicker than the others. Culm sheaths somewhat persistent, slightly pruinose. Sheath auricles falciform, setae yellowish-brown. Sheath ligules truncate at apex, greenish-yellow. Leaves 2 to 3 on twigs. Leaf sheaths glabrous, leaf auricles and setae obsolescent or not distinct. Leaf ligules slightly arcuate at apex. Leaf blades narrow-lanceolate, 11 to 18 cm long, 1.7 to 2.1 cm wide; puberulent beneath. Lateral veins 5 to 7 pairs. Transverse veinlets distinct.

Distributed in Yunhe Prefecture of Zhejiang Province. Tolerates light frost.

2. *Pleioblastus maculatus* (McClure) C. D. Chu et C. S. Chao

Rhizomes amphipodial. Culms 3 to 4 m tall, 1.5 to 2 cm in diameter. Culm wall 6 mm thick. Young culms green, densely fugacious-pruinose; glabrous except erectly to nearly retrorsely white-ciliolate under nodes only. Old culms yellowish-green, sparsely grayish-black powdery. Internodes 21 to 26 cm long in average, cylindrical, slightly grooved on branching side. Sheath scars with girdled brown cilia when young, glabrous when mature. Primary branches 3 to 5 on nodes. Leaves 3 to 5 on twigs. Leaf sheaths green and glabrous. Leaf blades lanceolate, 13.4 to 18.5 cm long, 2.3 to 2.9 cm wide; fugacious-tomentellate beneath, obtuse-cuneate at base, serrulate. Lateral veins (5) 6 (7) pairs.

Distributed in Guangxi and Sichuan Provinces. Cultivated as far north as in Shaanxi Province. Hardy to −15°C.

3. *Pleioblastus yixingensis* S. L. Chen et S. Y. Chen, sp. nov.

Rhizomes slender. Culms erect, 3 to 5 m tall, 1.2 to 2 cm in diameter; greenish-yellow, glabrous; densely pruinose, becoming to blackish farinose afterwards, somewhat green-yellowish. Internodes 17 to 18 cm long. Nodes prominently swollen. Intranodes 5 mm long. Primary branches 3 to 5 on nodes. Culm sheaths somewhat persistent, green, becoming pale yellowish-green later. Top of sheath proper semi-coriaceous, blackish-brown; dark-red ciliate on rims. Sheath blades lanceolate, spreading or reflexed. Leaves 3 to 5 on twigs. Leaf auricles variable in size and form, setae stramineous or purple, radiating. Leaf blades lanceolate, 13.5 to 20 cm long, 2 to 2.7 cm wide; green and glabrous above, pale green and scabrous beneath, white-pubescent along midrib; serrulate. Lateral veins 6 to 7 pairs.

Distributed in Hangzhou of Zhejiang Province. Hardy to −7°C.

4. *Pleioblastus rugatus* Wen et S. Y. Chen, sp. nov.

Culms 5 m tall, 2 cm in diameter. Internodes 35 cm long. Culm wall thick. Nodes slightly swollen, pruinose zone under nodes. Sheath scars white-pubescent. Leaves 3 to 4 on twigs. Leaf sheaths 5 cm long, glabrous, with transverse veinlets when young. Leaf ligules 2 mm long, ovate at apex, glabrous, pruinose. Leaf blades lanceolate to oblong, 11 to 18 cm long, 14 to 30 mm wide; obtuse-rounded at base, with a 2 to 3 mm long stalk; acute and prolonged at apex; glabrous in general. Lateral veins 5 to 7 pairs, tertiary veins 9 between pairs of lateral veins. Transverse veinlets exserted.

Distributed in Zhejiang Province. Hardy to −7°C.

5. *Pleioblastus hsienchuensis* Wen, sp. nov.

Culms 5 m tall, 2 to 3 cm in diameter. Internodes 30 cm long. Nodal ridges swollen; sheath scars with sheath callus, setose; pruinose under nodes. Leaves 4 to 5 on twigs. Leaf sheaths 4 mm long, glabrous, pruinose with longitudinal veins and transverse veinlets. Leaf auricles elliptic, protruding, with curved setae, 13 mm long. Tertiary veins 8 between pair of lateral veins.

Distributed in Zhejiang Province. Hardy to −7°C.

6. *Pleioblastus longifimbriatus* S. Y. Chen, sp. nov.

Rhizomes slender. Culms erect, 3 to 4 m tall, 1.5 cm in diameter; green, glabrous, sparsely purple-punctate and pruinose when young; dark green and blackish farinose afterwards. Culm pith sponge-like. Internodes 29 to 42 cm long. Nodes swollen. Culm sheaths persistent, semi-coriaceous, as long as ½ of internodes; green, lemon-colored-farinose later; glabrous on abaxial surface, except brown-ciliolate on rims. Sheath auricles dark green, narrow-falciform; setae purple, radiating, 1 to 1.5 cm long. Sheath ligules truncate at apex, 0.5 to 1 mm long, pubescent. Sheath blades lanceolate, pale green, reflexed. Leaves 4 to 6 on twigs. Leaf auricles green, variable in form. Leaf ligules slightly arcuate at apex. Leaf blades somewhat papery, elliptic-lanceolate, 10 to 13.5 cm long, 2 to 2.8 cm wide. Lateral veins 5 to 7 pairs.

Distributed in Hangzhou of Zhejiang Province. Hardy to −7°C.

7. *Pleioblastus juxianensis* T. H. Wen, C. Y. Yao, et. S. Y. Chen, sp. nov.

Rhizomes slender. Culms erect, 1.7 to 3 m tall, 1 to 3 cm in diameter; nearly solid, green; sparsely pruinose, but dense under nodes; glabrous. Internodes 20 to 28 cm long. Nodes swollen. Primary branches 5 on nodes. Culm sheaths somewhat persistent, green, pruinose; glabrous on upper part; brown-short-barbate on base, fugacious; ciliate on rims. Sheath auricles lunate, green, with distinct setae. Sheath ligules truncate at apex, pale green, pruinose afterwards. Sheath blades green, lanceolate, reflexed. Leaves 3 to 5 on twigs. Leaf sheaths glabrous. Leaf auricles lunate; setae blackish-brown. Leaf ligules arcuate, pruinose, 5 mm long, ciliolate on rims. Leaf blades lanceolate, 12 to 16 cm long, 2.3 to 2.6 cm wide. Lateral veins 6 to 7 pairs.

Distributed in Juxian Prefecture of Zhejiang Province. Hardy to −5°C.

8. *Pleioblastus kwangsiensis* W. Y. Hsiung et C. S. Chao, sp. nov.

Culms 5 m tall, 3 cm in diameter, cylindrical, not grooved. Young culms green, white-hirsute, not pruinose. Internodes on midculm 40 cm long. Culm pith stratum-like incrassate. Two girdles of hairs on sheath scars, the above girdle ascendant and the lower girdle retrorse. Primary branches 5 on nodes, slant-spreading. Leaves 3 to 4 on twigs. Leaf sheaths glabrous. Leaf blades lorate-lanceolate, 11 to 20 cm long, 1.3 to 1.8 cm wide.

Distributed in Guangxi Province. Tender.

9. *Pleioblastus oleosus* Wen, sp. nov.

Culms 3 to 5 m tall, 1 to 3 cm in diameter. Internodes 35 to 40 cm long, cylindrical, compressed from mid-internode downwards on branching nodes. Young culms pubescent, green, not pruinose. Old culms yellow, glabrous, blackish-dirty under nodes. Sheath scars narrow, not swollen. Nodal ridges swollen or slightly swollen. Primary branches 3 on nodes. Leaves 2 to 4 on twigs. Leaf sheaths 5.5 cm long, with longitudinal veins and transverse veinlets, glabrous, except ciliolate on rims. Leaf auricles and setae none. Leaf ligules truncate at apex, pubescent on rim of apex. Leaf blades lanceolate, 10 to 18 cm long, 13 to 22 mm wide, thin, acute at base, acuminate and apiculate at apex; glabrous, sharp-serrulate on one side of rim only. Lateral veins 4 to 5 pairs. Transverse veinlets obvious.

Distributed in Zhejiang, Fujian, Jiangxi and Yunnan Provinces. Hardy to −5°C.

10. *Pleioblastus altiligulatus* S. L. Chen et S. Y. Chen, sp. nov.

Rhizomes slender. Culms 2 to 3 m tall, 1 to 1.5 cm in diameter; green, pruinose, glabrous. Internodes 24 cm long. Nodes swollen. Primary branches 3 to 5 on nodes. Culm sheaths persistent, semi-coriaceous, green, glabrous except ciliate on rims. Sheath auricles none. Sheath ligules 3 mm long, bluish-farinose. Sheath blades lanceolate, drooping, green but pale purple on margins and at apex. Leaves 2 to 3 on twigs. Leaf sheaths persistent. Leaf auricles none. Leaf ligules 3 mm long. Leaf blades lanceolate, 12 to 14 cm long, 1.4 to 1.9 cm wide; green above, pale green beneath, densely pubescent. Lateral veins 5 to 6 pairs.

Distributed in Qingyuan Prefecture of Zhejiang Province. Hardy to −5°C.

11. *Pleioblastus intermedius* S. Y. Chen, sp. nov.

Rhizomes slender. Culms erect, 3 to 4 m tall, 1 to 2 cm in diameter; green, densely farinose, hairy; but becoming dark green or greenish-yellow, glabrous and blackish farinose when mature. Internodes 21 to 22 cm long. Nodes swollen. Primary branches 3 to 5 on nodes. Culm sheaths coriaceous, green; sparsely blackish-brown-ciliate and purple-hirsute on rims, somewhat persistent. Nodes of old culms ciliate after shedding of culm sheaths. Usually without sheath auricles and setae. Sheath ligules retuse or arcuate on apex, farinose on abaxial surface. Sheath blades lanceolate, spreading or slightly recurved. Leaves 3 to 4 (rarely 8) on twigs. Leaf sheaths persistent. Leaf auricles falciform, fugacious; setae radiating on rims, purple, 3 mm long. Leaf ligules membranous, mucro at apex, 5 mm long. Leaf blades lanceolate, 13 to 23 cm long, 2.5 to 3.2 cm wide; dark green and glabrous above, yellowish-green and white pubescent beneath. Lateral veins 7 to 8 pairs.

Distributed in Hangzhou of Zhejiang Province. Hardy to −7°C.

12. *Pleioblastus amarus* (Keng) Keng f.

Culms 3 to 4 m tall, 1.5 to 2 cm in diameter. Young culms pale green, pruinose. Old culms greenish-yellow, gray-mealy-maculate. Internodes 25 to 40 cm long, cylindrical, slightly grooved on branching side. Primary branches usually 5 on nodes. Leaves 3 to 4 (usually 3) on twigs. Leaf blades 8 to 20 cm long, 1.5 to 2.8 cm wide; dark green above, pale green beneath, white tomentellate. Lateral veins 6 to 7 pairs.

Distributed at an altitude under 1000 m in Yangtze Valley Provinces and Mount Qingling of

Pleioblastus amarus (King) Keng f.

Shaanxi Province. Hardy to −15°C.

Pleioblastus amarus var. *hangzhouensis* S. L. Chen et S. Y. Chen, var. nov.

Culms smooth, glabrous, not farinose. Culm sheaths green or purplish-green, lucid, not farinose, not maculate. Sheath auricles absent. Sheath blades linear-lanceolate.

Distributed in Hangzhou Province. Hardy to −7°C.

Pleioblastus amarus var. *subglabratus* S. Y. Chen, var. nov.

Culm sheaths glabrous, slightly farinose, but fugacious. Leaf blades reaching 26 cm long, 4.9 cm wide.

Distributed in Juxian Prefecture of Zhejiang Province. Hardy to −7°C.

Pleioblastus amarus var. *pendulifolius* S. Y. Chen, var. nov.

Leafy twigs drooping. Culm sheaths not farinose. Sheath ligules concave at apex.

Distributed in Hangzhou of Zhejiang Province. Hardy to −15°C.

Pleioblastus amarus var. *tubatus* Wen, var. nov.

Culms 3 m tall, 12 mm in diameter, green and glabrous. Nodal ridges swollen, but not ribbed. Leaves 2 to 3 on twigs. Sheath blades triangular, glabrous, with longitudinal veins. Leaf sheaths 55 mm long, semi-coriaceous, glabrous, with longitudinal veins, both margins of the apex slant-lifting, not ciliate. Leaf blades lanceolate, 9 to 17 cm long, 12 to 23 mm wide, serrulate or scabrous on both sides of rims; glabrous above, puberulent only along base of midrib beneath. Lateral veins 6 to 7 pairs. Transverse veinlets distinct.

Distributed in Zhejiang Province. Hardy to −7°C.

X. *Pseudosasa* Makino

Tall or shrub-like bamboos. Rhizomes amphipodial. Culms diffuse, nearly cylindrical. Nodal ridges not swollen. Primary branches 1 to 3 on nodes. Twig 1 only on branch nodes. Culm sheaths abscising late or persistent; hirsute or glabrous. Leaf oral setae not ciliate or curved-ciliate. Leaf blades long-lanceolate. Transverse veinlets exserted.

KEY to the Spp. of *Pseudosasa* Native to China

1a. Culms tall. Culm sheaths brown-setose on abaxial surface. 2

1b. Culms small, 1.2 to 6 m. tall, less than 2 cm in diameter. 3

2a. Culms 7 to 13 m tall, 5 to 6 cm in diameter. Sheath proper truncate at apex. Leaf ligules 1 to 2 mm long. Leaf blades serrulate on one side of rim only, the other side smooth; glabrous beneath. . . . . (1) *P. amabilis*

2b. Culms 8 m tall, 5 cm in diameter. Sheath proper broad and concave at apex. Leaf ligules prominent, 8 mm long. Leaf blades sharp-serrate; puberulent beneath. (2) *P. longiligula*

3a. Lateral veins and transverse veinlets indistinct on the under surface of leaf blades. Leaf blades narrow-oblong, setulose beneath. . . . . (3) *P. pubiflora*

3b. Lateral veins and transverse veinlets prominent on the under surface of leaf blades. 4

4a. Culm sheaths glabrous on abaxial surfaces. 5

4b. Culm sheaths hairy on abaxial surfaces. 7

5a. Leaf blades glabrous on under surfaces. 6

5b. Leaf blades densely pubescent. Sheath auricles very small. Leaf blades oblong-lanceolate. Lateral veins 5 to 6 pairs. . . . . (4) *P. subsolida*

6a. Sheath auricles half-moon-like. Sheath blades narrow-lanceolate. Leaf blades slender-lanceolate. Lateral veins 5 to 9 pairs. (5) *P. cantoni*

6b. Sheath auricles indistinct. Sheath blades broad-ovate-lanceolate. Leaf blades lanceolate to narrow-lanceolate. Lateral veins 6 to 7 pairs. . . . . (6) *P. gracilis*

7a. Culm sheaths pilosulose on abaxial surfaces. Leaf blades sometimes puberulent beneath. Lateral veins 5 to 8 pairs. . . . . (7) *P. orthotropa*

7b. Culm sheaths setose on abaxial surface, denser on base. Leaf blades glabrous. Lateral veins 6 to 7 pairs. . . . . (8) *P. aeria*

Prunus yedoensis, bamboos and a typical Chinese pavilion on the top of a man-made rockpile at a garden near the West Lake in Hanchou. (Fong)

Malus halliana in full bloom with bamboos as background. (Fong)

Plate 1

Plate 2

Bamboos in the Forbidden City. Along a wall and (opposite page) between two doorways.

Phyllostachys viridis 'Houseau' McClure. (Fong)

Phyllostachys bambusoides f. *tanake* Makino ex Tsuboi. (Fong)

Plate 4

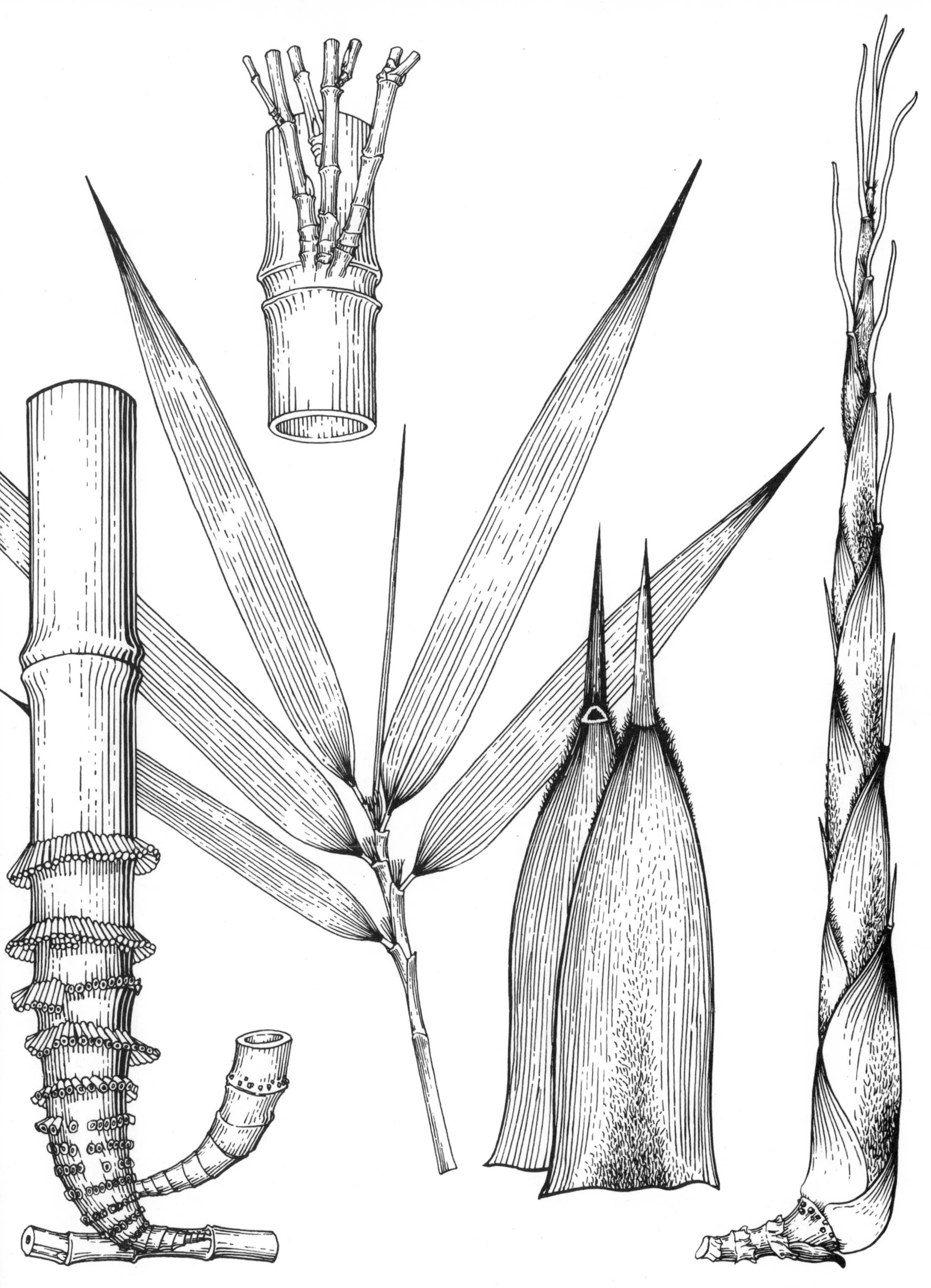

Pseudosasa amabilis (McClure) Keng f.

1. *Pseudosasa amabilis* (McClure) Keng f.
[*Arundinaria amabilis* McClure]

Culms 7 to 13 m tall, 5 to 6 cm wide; cylindrical, glabrous, pale green, pruinose. Primary branches usually 3 (rarely 1) on nodes, adnate. Culm sheaths late deciduous, dark brown, densely brown-setose, truncate at apex, with a bundle of oral setae on both sides. Sheath ligules brown, rounded, fimbriate. Sheath blades slender. Leaf sheaths glabrous except rims, with 2 bundles of oral setae. Leaf ligules 1 to 2 mm long, dense-lanuginose on rims. Leaf blades slender-lanceolate, thick, tensile; pale green and glabrous beneath; serrulate on one side of rim only, the other side smooth; slightly revolute. Lateral veins 7 to 9 pairs. Transverse veinlets exserted.

Distributed at low altitude in Guangdong, Guangxi and Hunan Provinces. Tender.

Pseudosasa amabilis var. *tenuis* S. L. Chen et G. Y. Sheng

Culm sheaths glabrescent, of thinner texture than the species. Sheath ligules shorter than the sp.

Distributed in Fujian Province. Tender.

2. *Pseudosasa longiligula* Wen, sp. nov.

Culms 8 m tall, 5 cm in diameter. Culms green glabrous, pruinose under nodes. Nodal ridges not swollen. Sheath callus on sheath scars. Primary branches 1 to 3 on nodes, adnate. Leaves 4 to 6 on twigs. Leaf sheaths ciliate. Leaf ligules strongly falcate. Leaf blades broad-lanceolate to narrow-lanceolate, 15 to 22 cm long, 13 to 14 mm wide; acute and prolonged at apex, acuminate or obtuse at base; green and glabrous above. Lateral veins 5 to 7 pairs.

Distributed in Guangxi Province. Tender.

3. *Pseudosasa pubiflora* (Keng) Keng f.

Culms 1.2 m tall, 2.5 mm in diameter, cylindrical. Primary branches 2 to 3 on nodes. Secondary branches bearing leaves. Leaf sheaths 3 to 4 on twigs, imbricate arranged, ciliate. Leaves 1 to 2 on tip of twigs. Leaf ligules tensile, truncate at apex. Leaf blades tensile, narrow-oblong; pale-grayish-green and setulose beneath; serrulate and scabrous on rim, or nearly smooth on lower rim.

Distributed in Guangdong Province. Tender.

4. *Pseudosasa subsolida* S. L. Chen et G. Y. Sheng, sp. nov.

Rhizomes slender. Culms nearly erect, 2.5 m tall, 5 to 12 mm in diameter, nearly solid, with 14 to 16 nodes. Internodes 18 to 30 cm long, cylindrical, slightly grooved near base. Nodes not swollen. Primary branches 1 to 3 on nodes. Culm sheaths papery, yellowish-brown, indistinctly maculate-striate, densely ciliate. Sheath ligules arcuate at apex. Sheath blades upright on upper-culm, recurved on lower-culm. Leaves 6 to 7 on twigs. Leaf blades 15 to 20 (23) cm long, 1.2 to 2.3 (2.7) cm wide; acuminate at apex, rounded at base; dark green and glabrous (except pilosulose near base) above, pale-greenish-yellow and dense-pilosulose beneath.

Distributed in Hunan Province. Tender.

5. *Pseudosasa cantoni* (Munro) Keng f.
[*Bambusa cantoni* Munro]

Culms 2 m tall, 5 mm in diameter. Culm lumina small. Nodal ridges not prominent. Culm sheaths late deciduous, as long as ½ of internodes, brownish-yellow, glabrous on abaxial surface, densely golden-yellow-ciliate on upper rims. Sheath auricles half-moon-like, with oral setae. Sheath ligules very short, truncate at apex. Sheath blades narrow-lanceolate. Leaf blades slender-lanceolate, long-acuminate and aristate at apex, dark green and lucid above, pale green and glabrous beneath. Lateral veins 5 to 9 pairs. Transverse veinlets prominent.

Distributed in Guangdong Province and Hong Kong. Tender.

6. *Pseudosasa gracilis* S. L. Chen et G. Y. Sheng, sp. nov.

Rhizomes slender. Culms nearly erect, 1.6 m tall, 4 mm in diameter; not pruinose under nodes, retrorse-pilose. Internodes 24 cm long, cylindrical. Nodes not swollen. Culm sheaths semi-coriaceous, abscising late. Primary branches 2 to 3 on nodes, 6 to 10 cm long. Leaves 2 to 3 on twigs. Leaf sheaths densely pubescent, densely ciliate. No leaf auricles, but oral setae remain. Leaf blades 14 to 19 cm long, 1.2 to 1.7 cm wide; acuminate at apex, cuneate at base, green and puberulent along midrib above.

Distributed in Hunan Province. Tender.

7. *Pseudosasa orthotropa* S. L. Chen et T. H. Wen, sp. nov.

Culms 4 m tall, 14 mm in diameter. Internodes 40 cm long, green and somewhat purplish, pruinose under nodes; white tomentose when young. Nodes not swollen. Sheath scars glabrous, not swollen or slightly swollen on upper part. Primary branches 1 to 3 on nodes, adnate at base and slightly spreading on top. Leaves 8 to 11 on twigs. Leaf sheaths ciliate. Leaf auricles triangular or oval, ciliate. Leaf ligules arcuate, puberulent and pruinose. Leaf blades long-acuminate at apex, 16 to 34 cm long, 15 to 35 mm wide; serrulate. Transverse veinlets exserted.

Distributed in Zhejiang and Fujian Provinces. Tender.

8. *Pseudosasa aeria* Wen, sp. nov.

Culms erect, 6 m tall, 2 cm in diameter. Internodes 30 to 40 cm long, green, glabrous. Nodes not swollen. Primary branches 1 to 3 on nodes. Leaves 3 to 5 on twigs. Leaf sheath glabrous except ciliate. Leaf auricles indistinct. Oral setae upright. Leaf ligules short, truncate at apex. Leaf blades 11 to 20 cm long, 12 to 20 mm wide; gradually narrowed at base to 3 mm long stalks; acuminate and caudate at apex.

Distributed in Zhejiang Province. Tender.

Appendix: *Pseudosasa japonica* (Sieb. & Zucc.) Makino

[*Arundinaria japonica* Sieb. & Zucc.]

Culms 2 to 4 m tall, 5 to 15 mm in diameter. Internodes green, glabrous, pruinose under nodes. Nodes not swollen. Branching from mid-culm, usually single on nodes. Leaf sheaths purplish at tip, with setae. Leaf blades slender-lanceolate, caudate-acuminate at apex, acute at base, glabrous, lucid above, whitish beneath, serrulate-minute-spiny on rims. Lateral veins 5 to 7 pairs, with transverse veinlets.

Distributed in Japan and Korea. Cultivated from the south bank of Yangtze River to Guangdong Province for ornamental use. Tolerant to −8°C in sheltered places in Shanghai.

A bamboo hedge in the South China Botanical Garden, Guangzhou.

XI. *Shibataea* Makino

Small shrub-like bamboos. Rhizomes amphipodial. Culms dwarf and thin, scattered or pluricespitose, somewhat zigzag. Nodes prominently swollen. Internodes smooth, compressed or half-cylindrical on branching side, nearly solid. Primary branches (3) 5 to 7 on nodes. Branches short (secondary branches none), usually 2 nodes, leaves 1 to 2 on upper nodes, with single membranous linear sheath on lower nodes. Culm sheaths papery or membranous, abscising early or late. Sheath auricles distinct. Sheath ligules triangular. Sheath blades usually short awn-form. Leaf blades thick papery, oblong to ovate-oblong; acute at apex; serrulate; glabrous above, lateral veins and transverse veinlets distinct beneath.

KEY to the Spp. of *Shibataea* Native to China

1a. Culms below 1 m tall, leaf blades glabrous beneath. 2
1b. Culms 1 to 2 m tall, leaf blades pilosulose beneath. 4
  2a. Culms about 60 cm tall, culm sheaths glabrous. .... (1) *S. chinensis*
  2b. Culms 50 cm tall, culm sheaths hairy. 3
    3a. Culms sheaths pale red; white-pubescent, especially dense on base; long-ciliate. Sheath scars puberulent. .... (2) *S. chiangshanensis*
    3b. Culm sheaths pale green, with thick, brown, shining, needle-like bristles; glabrous on base and rims. Sheath scars not hairy. .... (3) *S. strigosa*
      4a. Culm sheaths pale reddish-yellow, appressed-puberulent on abaxial surface. Primary branches 2 to 6 on nodes. .... (4) *S. kumasasa*
      4b. Culm sheaths dark brown, glabrous on abaxial surface. Primary branches 3 to 4 on nodes. .... (5) *S. hispida*

1. *Shibataea chinensis* Nakai

Culms 60 cm tall, 2 to 3 mm in diameter. Internodes 7 to 15 cm long, glabrous, pale green, slightly purplish. Nodal ridges swollen. Sheath callus none. Culm sheaths early deciduous, membranous, glabrous. A minimized sheath blade on top of sheath proper. No oral setae. Primary branches 3 to 6 on nodes. Leaf usually single on twig top. Leaf sheaths nearly coriaceous, glabrous. Leaf ligules membranous, usually oblique to one side, awl-shaped. Leaf blades thick-papery, ovate-lanceolate to broad-lanceolate; green and glossy above, glabrous beneath; serrulate, acuminate at apex, asymmetrically rounded at base. Transverse veinlets distinct.

Distributed in southern Anhui, Jiangsu, and Zhejiang Provinces. Cultivated for ornamental use, as ground covers, or pot-grown. Hardy to −10°C.

2. *Shibataea chiangshanensis* Wen, sp. nov.

Culms 0.5 m tall, 2 mm in diameter. Internodes 7 to 12 cm long, nearly half cylindrical, green when young, red-brownish when old. Primary branches 3 on nodes. The main branch two times longer than the others. Leaf single on twigs. Leaf stalk 8 mm long, excurrent from the top of twig. Usually no leaf sheaths. Leaf blades ovate to triangular, 6 to 8 cm long, 11 to 23 mm wide, the broadest part near base, blunt to truncate at base, acute and excurrent to caudate at apex; asymmetrical, especially on base; long-serrate; glabrous on both surfaces. Lateral veins 7 to 8 pairs.

Distributed in Zhejiang Province. Hardy to −7°C.

3. *Shibataea strigosa* Wen, sp. nov.

Culms 0.5 m tall, 3 mm in diameter. Internodes 13 cm long, usually not grooved on branching side, green and glabrous. Nodal ridges swollen, liratus. Primary branches 3 on nodes. The main branch as long as or slightly longer than the others. Leaf single on twigs. Leaf stalks usually 3 to 4 mm long. Leaf blades ovate-lanceolate to elliptic; blunt at base, acuminate and excurrent at apex; asymmetrical at base; 5 to 7 cm long, 1.5 to 2 cm wide, serrulate only on one side, smooth on the other side; glabrous on both surfaces. Lateral veins 6 to 7 pairs.

Distributed in Zhejiang Province. Hardy to −7°C.

4. *Shibataea kumasasa* (Zoll.) Nakai

Culms 1 to 2 m tall, 1 to 5 mm in diameter, glabrous and glossy. Nodal ridges swollen. Culm sheaths pale yellowish-red, papery; adnate; tomentellate on abaxial surfaces. Oral setae tubercle-based. Primary branches 2 to 6 on nodes. Leaf single on branch top. Leaf blades ovate-lanceolate, 3.5 to 7 cm long, 1 to 2 cm wide; acuminate at apex, rounded at base; green and glabrous above, grayish-white and pilosulose beneath. Lateral veins

Shibataea chinensis Nakai

6 to 9 pairs. Transverse veinlets not distinct.

Distributed in Fujian and Taiwan Provinces. Ornamental. Tender.

5. *Shibataea hispida* McClure

Culms a little over 1 m tall, 1.5 to 4 mm wide. Internodes 8 to 19 cm long, glabrous and glossy. Nodal ridges prominently swollen. Culm sheaths late deciduous, dark brown, glabrous and lineate on abaxial surfaces. Sheath auricles and ligules none. Sheath blades very small. Primary branches 3 to 4 on nodes. Usually single leaf on branch top. Leaf sheaths very short, glabrous or pilosulose near base. Leaf blades ovate lanceolate, 7.5 to 10 cm long, 2 to 3 cm wide; acuminate at apex, rounded at base; green and glabrous above, grayish green and with short white bristles beneath. Lateral veins 6 to 8 pairs. Transverse veinlets distinct.

Distributed in southern Anhui Province. Cultivated as an ornamental. Hardy to −7°C.

XII. *Gelidocalamus* Wen, gen. nov.

Shrub-like bamboos. Rhizomes amphipodial. Internodes cylindrical, not grooved. Primary branches 7 to 12 clustered on nodes, slender, bearing 2 to 3 nodes only, no secondary branching. Leaf one on top of branch. Branch sheaths longer than internodes. Bamboo shoots appear in winter. Culm sheaths persistent. Sheath auricles obsolescent or obsolete. Sheath ligules extremely short, arcuate or truncate at apex. Sheath blades short-subulate. Leaf blades lanceolate to broad-lanceolate, acute and caudate at apex. Transverse veinlets visible on both surfaces.

KEY to the Spp. of *Gelidocalamus* Native to China

1a. Sheath auricles very small. Pruinose zone under nodes of young culms. Culm sheaths glabrous on abaxial surface. Leaf blades smaller, 12 to 17 cm long, 13 to 22 mm wide. .... (1) *G. stellatus*
1b. Sheath auricles absent 2
  2a. Young culms densely white-tomentose only under nodes. Culm sheaths purplish-brown-square-maculate and sparse-hairy on abaxial surface. Leaf blades 19 to 23 cm long, 24 to 32 mm wide. .... (2) *G. tessellatus*
  2b. Young culms densely white-lanose, not pruinose but hairy girdle under nodes. Culm sheaths pale-red, hirsute. Leaf blades large, 18 to 27 (31) cm long, 22 to 30 (43) mm wide. . . . (3) *G. rutilans*

1. *Gelidocalamus stellatus* Wen, sp. nov.

Culms 2 m tall, 8 mm in diameter, green and glabrous when young. Internodes 25 to 30 cm long, cylindrical, not grooved, pruinose under nodes. Nodes not swollen. Primary branches 7 to 12 on nodes of nearly even thickness, but not equal in length. Leaf sheath usually none. Branch sheaths overlap the tip of branch. Leaf blades gradually narrowed or obtuse and contracted to stalk, serrulate and scabrous on one side only, (the other side smooth), green and glabrous above, powdery-green and puberulent along base of midrib only on the under surface. Lateral veins 4 to 5 pairs, tertiary veins 7 between every pair of lateral veins.

Distributed in Mount Jingganshan of Jiangxi Province. Ornamental. Hardy to −5°C.

2. *Gelidocalamus tessellatus* Wen et C. C. Chang, sp. nov.

Culms over 3 m tall, 1 cm in diameter. Internodes 30 cm long; sparsely hirsute when old. Intranodes glabrous. Nodal ridges slightly swollen. Culm sheaths coriaceous, ciliate on rim; white-tomentulose on top. Primary branches 12 on nodes of even thickness but not equal in length, with nearly persistent branch sheaths. The top of branch sheaths overlap the tip of branches. Leaf blades broad-lanceolate, obtuse to acuminate at base; smooth on both sides of rim or serrulate on one side only; green and glabrous above; powdery-green and puberulent along midrib, especially at base on under surface. Lateral veins 7 pairs, tertiary veins 6 to 7 between every pair of lateral veins.

Distributed in Guizhou Province. Hardy to −5°C.

3. *Gelidocalamus rutilans* Wen, sp. nov.

Culms 1 m tall, 3 to 6 mm in diameter. Internodes 10 to 15 cm long. Hairy girdle under nodes. Nodal ridges slightly swollen. Culm sheaths 20 cm long, glabrous on margins, truncate or obtuse-roundish at apex. Primary branches 3 to 8 on nodes, slender, nearly equal in size. Branch sheaths persistent or late deciduous. Leaf sheaths usually obscure or indistinct. Leaf blades acute or obtuse at base and contracted to short stalk, glabrous above, hirsute only near base beneath; serrulate. Lateral veins 6 to 8 pairs. Transverse veinlets visible beneath.

Distributed in Zhejiang Province. Hardy to −7°C.

Gelidocalamus stellatus Wen

XIII. *Phyllostachys* Siebold et Zuccarini

Giant or shrub-like bamboos. Rhizomes monopodial, culms diffuse. Internodes compressed or grooved on branching side. Culm sheaths abscising early, usually maculate. Sheath blades narrow and long. Sheath auricles and oral setae prominent or absent, sheath ligules exserted. Primary branches 2 on nodes; an early deciduous scale, 2 lobed or entire, on the base of primary branches; but no such scale on the base of any secondary branches. Leaves 1 to several on twigs (sometimes up to 10 on young culms). Leaf blades linear-lanceolate to lanceolate, midrib prominent, with transverse veinlets.

KEY to the Spp. of *Phyllostachys* Native to China

1a. Culm sheaths maculate on abaxial surfaces. Sheath blades usually narrower than sheath ligules, linear, (but often broader on the type with diffuse small blotches), reflexed or spreading, rarely upright; heavily plicate to smooth, usually spread on the top of bamboo shoots. Sheath ligules over 3 mm long. Intranodes usually 3 mm wide. 2
1b. Culm sheaths not maculate on abaxial surfaces. The base of sheath blade at least as wide as ½ of sheath ligule, or nearly as wide; more or less triangular, upright, flat or curved, but seldom plicate; usually appressed together on the top of bamboo shoots. Sheath ligules usually 2 mm long. Intranodes usually 5 mm wide. 29
2a. Sheath auricles and oral setae none. Culm sheaths usually glabrous on abaxial surfaces. 3
2b. Sheath auricles remain, or obsolescent but with 10 mm long oral setae. Culm sheaths more or less bristly on abaxial surfaces. 17
3a. Culm surface, when seen under magnifying glass, scattered with minute granules of white crystals or minute pores, more prominent on upper part of internodes. 4
3b. Culm glabrous, without minute granules or minute pores. 5
4a. Nodal ridges of unbranching lower-culm nodes obsolescent or less swollen than sheath scars. Sheath ligules pale-green- or white-ciliate when young. .... (1) *Ph. viridis*
4b. Nodal ridges of unbranched lower-culm nodes prominent or as swollen as sheath scars. Sheath ligules purplish-red-ciliate when young. .... (2) *Ph. makinoi*
5a. On young culms, sheath scars and the base of culm sheaths white-ciliate or puberulent. 6
5b. On young culms, sheath scars and the base of culm sheaths glabrous. 7
6a. Internodes on lower-culm (sometimes up to mid-culm) often extremely shortened, nodes crowded, thickened and asymmetrically swollen. Sheath ligules ciliate, cilia longer than the ligule. .... (3) *Ph. aurea*
6b. Culms normal. Sheath ligules ciliate, cilia shorter than the ligules. .... (4) *Ph. meyeri*
7a. Sheath ligules ciliate, cilia long, purplish-red, dark brown or white. 8
7b. Sheath ligules white-ciliolate. 10
8a. Sheath ligules broad, arcuate at apex, long-purplish-red-ciliate. 1 to 2-year-old culms yellowish-green-striate (barely visible) from mid-culm downward. .... (5) *Ph. iridenscens*
8b. Sheath ligules narrow, truncate at apex, with long, white or dark brown cilia. Culms not striate. 9
9a. Culms sheaths pale yellow or pale yellowish-green, sparsely maculate. Sheath ligules pale yellowish-green, long-grayish-ciliate. .... (6) *Ph. angusta*
9b. Culm sheaths usually of darker shade, densely maculate. Sheath ligules dark brown or greenish-brown, with long, thick, brown cilia. .... (7) *Ph. flexuosa*
10a. Sheath blades smooth or slightly wrinkled. Sheath ligules truncate or arcuate at apex, rarely falcate and decurrent on both corners as in those spp. with young culms striate and culm sheaths verrucate-setose. 11

10b. Sheath blades intensely plicate. Sheath ligules highly falcate or sometimes arcuate at apex, decurrent or extending horizontally sidewards from both corners. 15

11a. Young culms not purplish-maculate, or both young and old culms maculate. Culm sheaths not scabrous, sometimes sparsely hispidulous. 12

11b. Lower part of young culms purple-maculate. Mature culms (except *Ph. nuda* f. *localis*) not maculate. 14

12a. Culms not or slightly pruinose. Culm sheaths green, densely maculate of unequal sizes. Sheath ligules greenish-purple, convex at center of apex, decurrent on both corners. . . . . (8) *Ph. acuta*

12b. Young culms pruinose, sometimes pruinose zone only under nodes. Culm sheaths pale reddish-brown, yellowish-brown, or sometimes greenish. Sheath ligules truncate or arcuate at apex, not decurrent. . . . 13

13a. Culm sheaths not pruinose. Sheath ligules truncate at apex, dark purple. . . . (9) *Ph. glauca*

13b. Culm sheaths pruinose. Sheath ligules arcuate at apex, pale brown. . . . . (10) *Ph. propinqua*

14a. Leaf blades villous on the base beneath. Sheath ligules 4 mm long, truncate or arcuate at apex, not decurrent. . . . . (11) *Ph. nuda*

14b. Leaf blades glabrous or glabrescent on the base beneath. Sheath ligules 4 to 8 mm long, highly falcate at apex, bow-shaped, decurrent on both corners. . . . . . (12) *Ph. arcana*

15a. Young culms not pruinose. Sheath ligules arcuate at apex, undulate on rim. . . . . (13) *Ph. glabrata*

15b. Young culms pruinose. Sheath ligules highly falcate at apex, decurrent. . . . 16

16a. The longest internodes on mid-culm over 25 cm long. Young culms green, slightly pruinose. Nodes not purple. . . . . (14) *Ph. virax*

16b. The longest internodes on mid-culm 20 to 25 cm long. Young culms densely pruinose. Nodes purple. . . . . . . (15) *Ph. praecox*

17a. Sheath auricles small or obsolescent, but long oral setae exserted. Sheath ligules fimbriate-ciliate, cilia 5 mm long. 18

17b. Sheath auricles prominent, long, usually falciform. Sheath ligules ciliolate. 20

18a. Nodal ridges of unbranching lower-culm nodes not swollen. Sheath ligules highly falcate at apex. . . . . (16) *Ph. pubescens*

18b. Nodal ridges of unbranching lower-culm nodes swollen. Sheath ligules truncate to arcuate at apex. 19

19a. Intranodes of young culms densely pubescent. Culm sheaths purplish-brown. . . . . (17) *Ph. kwangsiensis*

19b. Intranodes of young culms glabrous or glabrescent. Culm sheaths pale greenish-yellow. . . . . (18) *Ph. circumpilis*

20a. Sheath scars of young culm hairy. Sheath ligules highly falcate at apex. . . . (19) *Ph. nigra*

20b. Sheath scars of young culms glabrous. Sheath ligules arcuate at apex. 21

21a. Culm sheaths glabrous on abaxial surface. Sheath auricles obviously coherent to sheath blades, (except in *Ph. decora,* sheath auricles small, not obviously coherent to sheath blades). 22

21b. Culm sheaths more or less hairy on abaxial surface. Sheath auricles not coherent to sheath blades. 23

22a. Sheath ligules ciliate, cilia shorter than ligule. Young culms pruinose. . . . . (20) *Ph. aureosulcata*

22b. Sheath ligules ciliate, cilia longer than ligule. Young culms not or slightly pruinose . . . . (21) *Ph. decora*

23a. The length of sheath ligule equal to or less than 1/6 of its breadth, highly falcate at its apex, long-ciliate. 24

23b. The length of sheath ligule more than 1/6 of its breadth, sometimes decurrent on both sides. 25

24a. Sheath ligule somewhat highly falcate at apex, culms not prominent-ribbed-striate. . . . . (22) *Ph. viridi-glaucescens*

24b. Sheath ligules arcuate at apex, culms prominent-ribbed-striate. . . . . (23) *Ph. elegans*

25a. Young culms not or sparse-pruinose, or pruinose zone only below nodes. Nodal ridges highly swollen to slightly swollen. 26

25b. Young culms prominently pruinose. Nodal ridges slightly swollen. 28

26a. Sheath blades smooth or slightly wrinkled. Sheath auricles small. Nodal ridges slightly swollen. . . . . (24) *Ph. bambusoides*

26b. Sheath blades heavily corrugate. Sheath auricles usually large, curved falciform. Nodal ridges more or less prominently swollen. 27

27a. Sheath auricles green. Sheath ligules pale brown, ciliate. Nodal ridges swollen. . . . . (25) *Ph. dulcis*

27b. Sheath auricles purplish-brown or blackish-purple. Sheath ligules long-ciliate. Nodal ridges highly swollen. . . . . (26) *Ph. prominens*

28a. Sheath ligules long-ciliate. Young culms green. Intranodes short. . . . . (27) *Ph. yunhoensis*

28b. Sheath ligules short-ciliate. Young culms purple. Intranodes 5 mm long. . . . . (28) *Ph. platyglossa*

29a. Sheath auricles broad and large, falciform or triangular; in case sheath auricles not large, prominent hairy zones on sheath scars of young culms. 30

29b. Sheath auricles none or small. 32

30a. Culm sheaths not striate. Sheath auricles curving-falciform or oblong. Sheath ligules falcate in the middle, long-ciliate. Leaves 2 to 3 on twigs. . . . . (19) *Ph. nigra*

30b. Culm sheaths striate. 31

31a. Culm sheaths pale-yellow-striate, somewhat pubescent, but denser on nodes. Sheath auricles broad and large, triangular. Leaf 1 (rarely 2) on twigs. . . . . (29) *Ph. nidularia*

31b. Culm sheaths purplish-yellow, purple-striate, sparsely brown-pricky. Sheath auricles falciform on upper culm sheaths, but small on lower culm. Leaves 2 on twigs. . . . . . . (30) *Ph. guizhouensis*

32a. Sheath auricles small, or exserted only in upper culm. 33

32b. Sheath auricles absent or obsolescent. 36

33a. Sheath auricles prominent, with a few short cilia. Sheath ligules short-ciliate. 34

33b. Sheath auricles obsolescent or absent, but oral setae long, 0.5 cm or more. Sheath ligules ciliate, 5 mm long. .... (31) *Ph. robustiramea*

34a. Culm sheaths purple-striate and bristly (glabrate later) on abaxial surface. Bristles more dense on those near base of culm. .... (32) *Ph. stimulosa*

34b. Culm sheath not purple-striate, glabrous or rarely sparse bristly. 35

35a. Young culms glabrous. The base of sheath blades as wide as the breadth of sheath ligules. .... (33) *Ph. heteroclada*

35b. Upper part of young culms hairy. The base of sheath blades as wide as 1/2 to 1/3 of the breadth of sheath ligules. .... (34) *Ph. bissetii*

36a. Culm sheaths bright red, pale purple or brownish-purple at earlier stage. Leaf blades small, 3 to 6.5 cm. long; but a little longer in the case of *Ph. rubicunda*. 37

36b. Culm sheath pale green to dark green at earlier stage. Leaf blade larger. 39

37a. Sheath ligules truncate or slightly arcuate at apex. Leaf stalks short, 1.5 to 2.5 mm long. 38

37b. Sheath ligules arcuate or falcate at apex. Leaf stalks longer and more slender, (2) 3 to 5 mm long. .... (35) *Ph. parvifolia*

38a. Upper leaf sheaths dense-villous. Leaves 3 to 5 on twigs. Sheath ligules long-ciliate. .... (36) *Ph. rivalis*

38b. Leaf sheaths ciliate only, or sparse-hairy. Leaves 2 to 3 on twigs. Sheath ligule cilia very short. ..... (37) *Ph. rubicunda*

39a. Culm sheaths blackish-green, somewhat purplish. Sheath ligule cilia minute, seen only under magnifying glass. .... (38) *Ph. atrovaginata*

39b. Culm sheaths pale green to green. Sheath ligule cilia short, but visible. .... (39) *Ph. concava*

1. *Phyllostachys viridis* (Young) McClure

Culms 10 to 15 m tall, 4 to 10 cm in diameter, green. Internodes on midculm 20 to 45 cm long. Young culms glabrous, slightly pruinose. Old culms pruinose zone only under nodes. Nodal ridges on nodes below branching nodes not prominent, but sheath scars swollen. Leaves 2 to 6 on twigs. Leaf blades lorate-lanceolate to lanceolate, 6 to 16 cm long, 1 to 2.2 cm wide.

Distributed in central China up to the Yellow River. Hardy to −20°C.

Phyllostachys viridis f. *houzeauana* C. D. Chu et C. S. Chao

The longitudinal grooves on the internodes pale yellow. Cultivated as an ornamental. Hardy to −20°C.

Phyllostachys viridis f. *youngii* C. D. Chu et C. S. Chao

Culms usually smaller, golden yellow; green zone under node. A few green longitudinal stripes on internodes. Culm sheaths yellowish-green, sparsely brown-maculate. Leaf blades pale-yellow-striate.

Distributed in Zhejiang Province. Hardy to −20°C.

2. *Phyllostachys makinoi* Hayata

This sp. is similar to *Ph. viridis*, but nodal ridges prominent on nodes below the branching nodes, or sometimes as swollen as the sheath scars. Sheath ligules purple, purple-ciliate on the apex.

Distributed in Fujian Province. Tender.

3. *Phyllostachys aurea* Carr. ex A. et C. Riviere

Culms 5 to 8 m tall, 2 to 3 cm in diameter, green. Old culms yellowish-green or grayish-green. Internodes on the lower culm, or sometimes up to midculm, extremely shortened and asymmetrically swollen. Sheath scars and the base of culm sheaths white-ciliate. Leaves 2 to 3 on twigs. Leaf blades lorate-lanceolate to lanceolate, 6 to 12 cm long, 1 to 1.8 cm wide.

Cultivated in Yangtze River Valley. Wild plants still exist in Zhejiang and Fujian Provinces. One of the most important garden bamboos. Hardy to −20°C.

4. *Phyllostachys meyeri* McClure

Culms 5 to 8 m tall, 2 to 3 cm in diameter. The base of culm sheath and the sheath scars on young

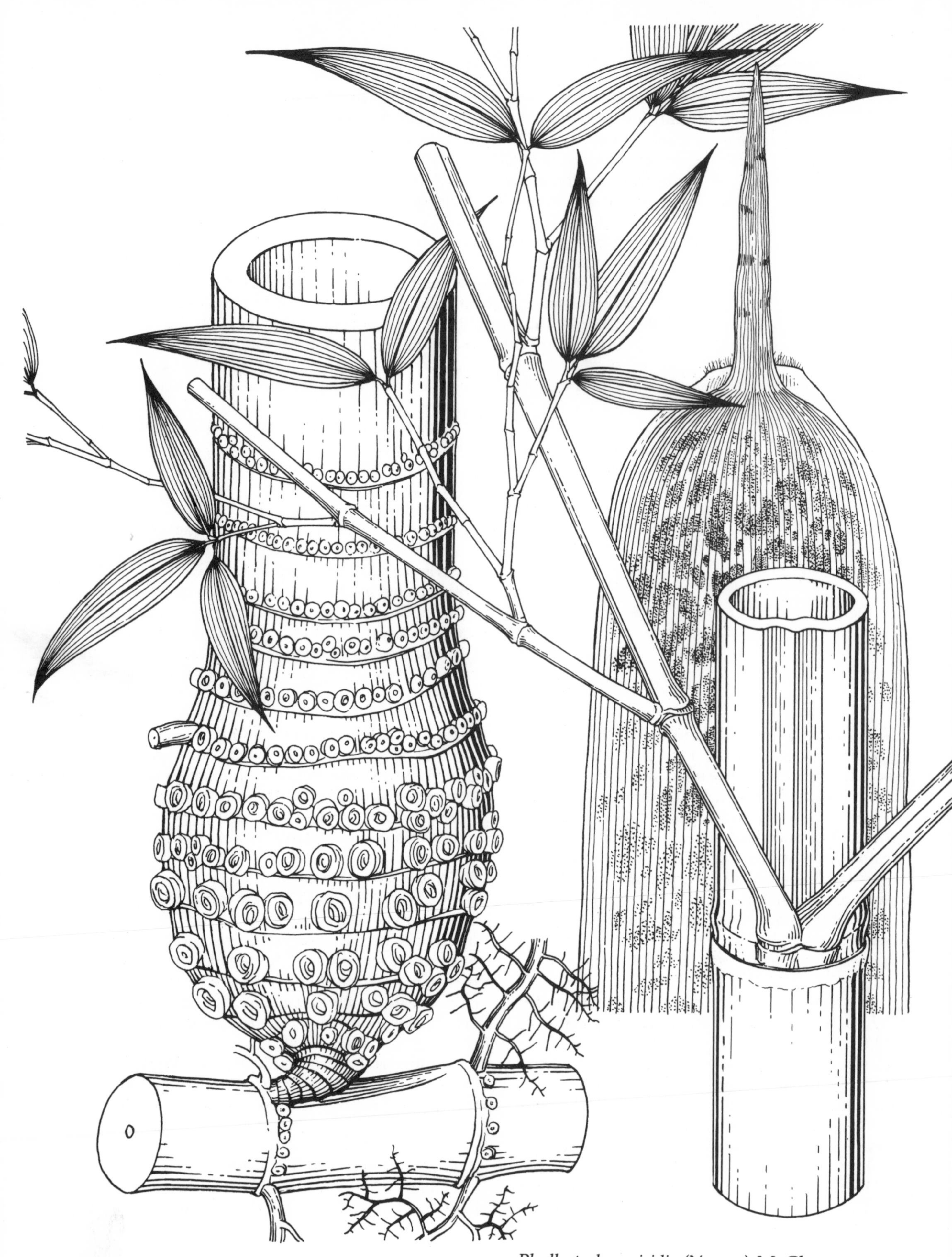

Phyllostachys viridis (Young) McClure

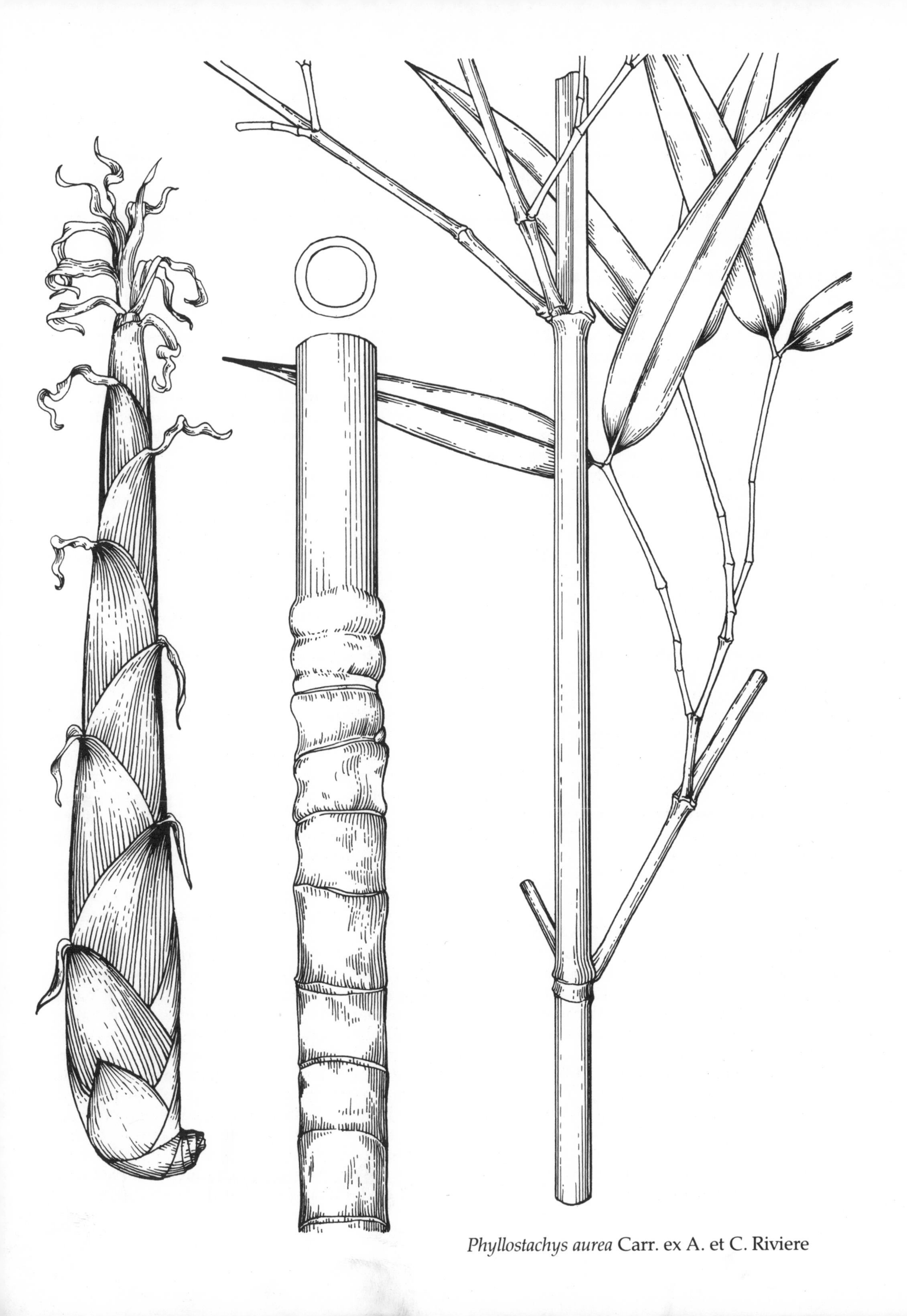

Phyllostachys aurea Carr. ex A. et C. Riviere

culms white-ciliate, somewhat similar to *Ph. aurea,* but all the internodes of culms normal, not shortened. Culm sheaths darker in color, more densely maculate. Sheath ligules longer, protruding in center of the apex, short-ciliate.

Distributed in Zhejiang, Anhui, and Hunan Provinces. Hardy to −7°C.

5. ***Phyllostachys iridenscens*** C. Y. Yao et S. Y. Chen, sp. nov.

Culms 10 to 12 m tall, 6 to 7 cm in diameter. The longest internodes on midculm up to 32 cm long. Nodal ridges slightly swollen. Young culms green, slightly pruinose; bluish-gray-farinose under nodes. Nodes on lower culm purplish. Old culms green to yellowish-green, usually yellow-striate. Culm sheaths pale-reddish-brown, darker on the margins and the apex, glabrous, densely purplish-black-maculate. Sheath ligules dark purple. Leaves 2 to 4 on twigs. Leaf sheaths purplish on the margins. Leaf auricles none, but purplish-red oral setae exist when young. Leaf ligules purple. Leaf blades 8 to 13 cm long, 1.2 to 1.8 cm wide, puberulent on the base beneath.

Distributed in Zhejiang and Jiangsu Provinces. Ornamental. Hardy to −10°C.

6. ***Phyllostachys angusta*** McClure

Culms 7.5 m tall, 4 cm in diameter, lateral branches short, crown narrow, conical. Internodes on midculm up to 26 cm long. Young culms powdery green; pruinose, especially below nodes. Old culms grayish-green. Nodal ridges slightly swollen. Culm sheath creamy-white or yellowish-green, sparsely maculate. Sheath ligules pale-yellowish-green, lacerate at apex. Leaves 2 (rarely 1) on twigs. Leaf blades 8.5 to 15 cm long, 1.3 to 2 cm wide, white-hairy on the base beneath.

Distributed in Jiangsu and Zhejiang Provinces. Hardy to −10°C.

7. ***Phyllostachys flexuosa*** A. et C. Riviere

This sp. is similar to *Ph. angusta;* but a few nodes on the base of culms zigzag; culm sheath greenish-yellow-brown, densely maculate, (blotches of unequal sizes); sheath ligules dark chestnut-red or yellowish-green; dark-colored-thick and long-ciliate at apex.

Cultivated plant. Hardy to −10°C.

8. ***Phyllostachys acuta*** C. D. Chu et C. S. Chao, sp. nov.

Culms 8 m tall, 4 to 6 cm in diameter. Intranodes contracted. Internodes on midculm 20 to 25 cm long, not pruinose. Young culms dark green, nodes purple. Old culms green or yellowish-green. Nodal ridges slightly swollen. Leaves 3 to 5 on twigs. Leaf blades lorate-lanceolate to lanceolate, 9 to 17 cm long, 1 to 2.2 cm wide, puberulent beneath, more densely-strewn along veins.

Distributed in Zhejiang Province. Hardy to −7°C.

9. ***Phyllostachys glauca*** McClure

Culms 10 to 12 m tall, 2 to 5 cm or more in diameter. Culm tops slightly arching. Internodes on midculm 30 to 40 cm long. Young culms pruinose, bluish-green, glabrous. Old culms green or grayish-yellow-green, pruinose zone only under nodes. Nodal ridges and sheath scars swollen. Leaves 2 to 3 on twigs. Leaf blades lorate-lanceolate to lanceolate, 8 to 16 cm long, 1.2 to 2.4 cm wide, hairy near base.

Distributed in the middle and lower reaches of Yellow River and Yangtze River. Can grow in hilly areas, plains, and flood lands, stands some dryness, poor soil; tolerates a slightly alkaline soil and −18°C, grows as far north as 40°N in northeastern China.

Phyllostachys glauca f. *yuozhu* J. L Lu, f. nov.

Culms gradually purplish-brown-maculate or striate.

Distributed in Hunan and Shaanxi Provinces. Ornamental. Hardy to −18°C.

10. ***Phyllostachys propinqua*** McClure

Culms 10 m tall, 5 cm in diameter. Internodes of midculm 26 to 38 cm long. Young culms bright green, pruinose zone prominent under nodes; sometimes internodes appearing bluish-green owing to pruinose. Nodal ridges slightly swollen. Leaves 2 to 3 on twigs. Leaf blades 7 to 16 cm long, 1.3 to 2 cm wide, hairy on the base beneath.

Distributed in Zhejiang, Jiangsu, Guangxi, Hunan, and Anhui Provinces. Hardy to −18°C. Can grow in Beijing gardens with some shelter.

11. ***Phyllostachys nuda*** McClure

Culms 8 m tall, 3 to 4 cm in diameter. The longest internode on midculm 30 cm long. Young culms dark green, dense-pruinose, sticky. Nodes purple. Intranodes purple-striate. Old culms green to grayish-green. Pruinose zone under nodes prominent. Culm ridges swollen and protruding. Culm sheaths pale-brownish-purple or pale-reddish-brown, coarse, pruinose, purple-multi-striate, maculate with large blotches, glabrous. Sheath ligules long. Sheath blades green, purple-lineate. Leaves 4 on twigs, only two (rarely one) left later. Leaf blades lorate-lanceolate to lanceolate, 8 to 10

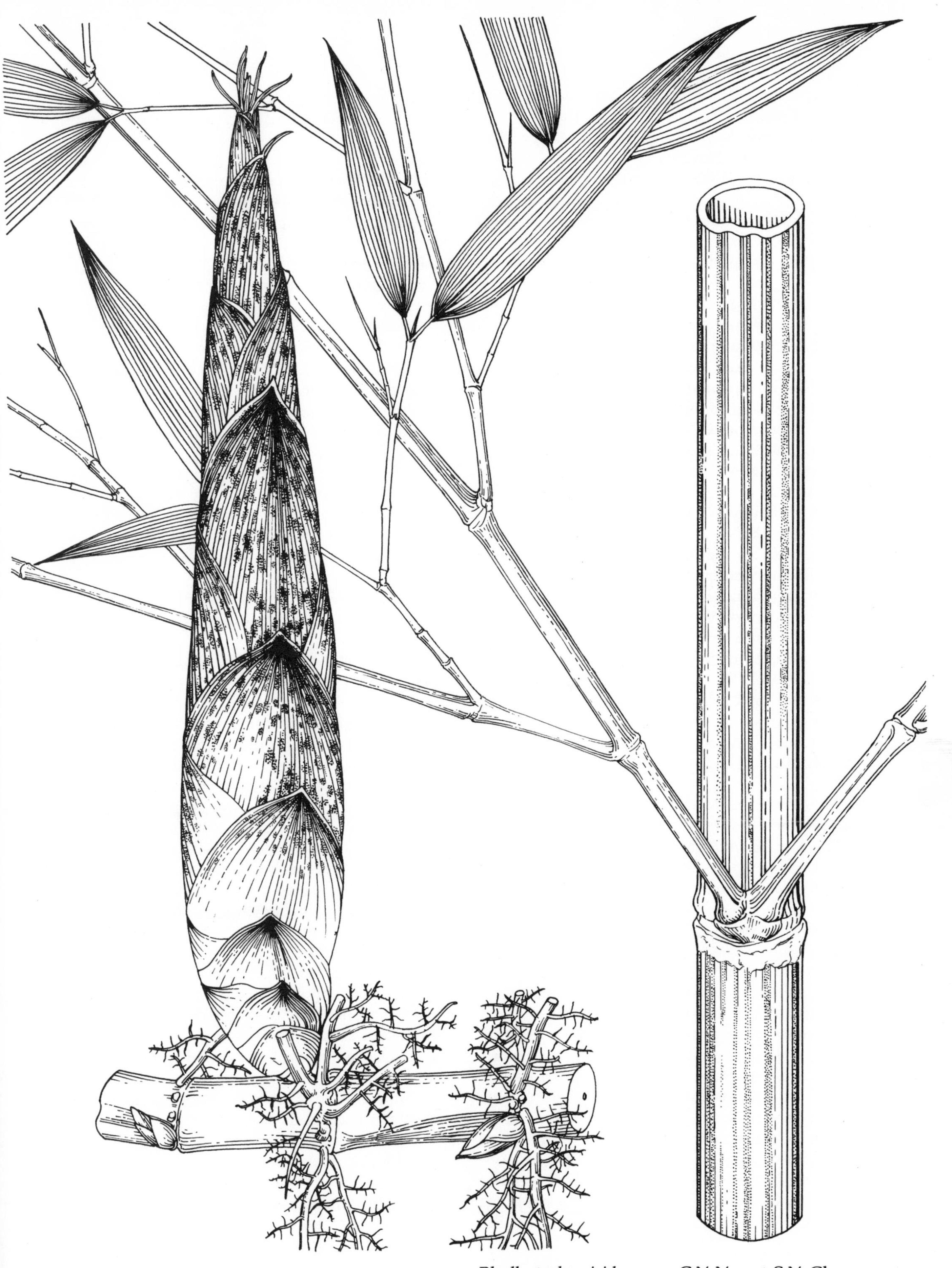

Phyllostachys iridenscens C.Y. Yao et S.Y. Chen

cm long, 1.5 to 2 cm wide, of thin texture, hairy near the base beneath.

Distributed in Zhejiang, Jiangsu, Anhui, and Hunan Provinces. Hardy to −12°C.

Phyllostachys nuda f. *localis* C. P. Wang et Z. H. Yu, f. nov.

Several internodes on culm base maculate with purplish-brown blotches, sometimes so dense as to make the internodes look purplish-brown in whole.

Distributed in Zhejiang Province.

12. *Phyllostachys arcana* McClure

Culms 8 m tall, 3 cm in diameter. Some culms zigzag. The longest internodes on midculm 25 cm long. Young culms bright green, slightly pruinose, thicker under nodes. Nodes purple, lower internodes purple-maculate. Old culms green or yellowish-green. Nodal ridges protruding and swollen, especially on the branching side. Intra-nodes broad and shallow. Culm sheaths yellowish-green or greenish, sometimes pale purplish on the lower part; margins orange-yellow; pruinose, scabrid, purple-lineate, scattered with small purplish-black spots. Upper culm sheaths sometimes not wholly maculate except on the base of culm sheaths only. Leaf auricles and oral setae none. Sheath blades lorate, green, purple-lineate, smooth, reflexed. Leaves 2 (rarely 1) on twigs. Leaf blades lorate-lanceolate, 7 to 11 cm long, 1.1 to 1.6 cm wide, glabrous.

Distributed in Jiangsu, Zhejiang, Anhui, Sichuan, Shaanxi, and Gansu Provinces. Hardy to −20°C.

Phyllostachys arcana f. *luteosulcata* C. D. Chu et C. S. Chao, f. nov.

The longitudinal grooves on culms yellow in color.

Distributed in Jiangsu Province. Ornamental.

13. *Phyllostachys glabrata* S. Y Chen et C. Y Yao, sp. nov.

Culms 6 to 7 m tall, 3 to 4 cm in diameter. The longest internodes on midculm 19 cm long. Young culms dark green, not pruinose, glabrous, scabrid, no pruinose zone under nodes. 2 to 3-year-old culms grayish-green. Nodal ridges as swollen as sheath scars. Culm sheath thin, pale reddish-brown to pale purplish-yellow; densely strewn with purplish-brown spots, aggregated into large blotches at upper part; not pruinose, glabrous, smooth. Sheath blades lorate, reflexed rugose, purplish green, margins purplish-red to orange-yellow. Leaves 2 to 4 on twigs. Leaf blades 8 to 11 cm long, 1.2 to 2 cm wide. Leaf auricles green, densely green-ciliate.

Distributed in Zhejiang Province. Hardy to −7°C.

14. *Phyllostachys virax* McClure

Culms 10 to 15 m tall, 4 to 8 cm in diameter. Culm tops arching. Old culms grayish-green or yellowish-green; pruinose zones under nodes; slender-longitudinal-liratus. Culm nodes oblique, nodal ridges usually asymmetrically swollen. Culm sheaths pale yellow, densely-blackish-brown-maculate (with spots and blotches). Leaves 2 to 4 on twigs. Leaf blades lorate-lanceolate to lanceolate, 9 to 18 cm long, 1.1 to 1.5 (2) cm wide, dark green, slightly drooping, with tufted hairs or nearly glabrous on the base beneath.

Distributed in Zhejiang, Jiangsu, and Hunan Provinces. Hardy to −20°C.

Phyllostachys virax f. *huanwenzhu* J. L Lu, f. nov.

The longitudinal grooves on culms yellow-striate.

Distributed in Hunan Province. Hardy to −23°C. Ornamental.

Phyllostachys virax f. *vittata* Wen, f. nov.

Internodes without pruinose zone. Culm sheaths dark red; longitudinal-brown-striate.

15. *Phyllostachys praecox* C. D. Chu et C. S. Chao, sp. nov.

Culms 8 to 10 m tall, 4 to 6 cm in diameter. Internodes on midculm 15 to 25 cm long, asymmetrically swollen. Young culms dark green, nodes purplish-brown, glabrous. Old culms yellowish-green, or grayish-green, sometimes not distinctly yellow-striate. Nodal ridges and sheath scars moderately swollen. Leaves 2 to 3 (rarely 5 to 6) on twigs. Leaf blades linear-lanceolate (the lowest leaf on each twig shorter, lanceolate), 6 to 18 cm long, 0.8 to 2.2 cm wide.

Distributed in Zhejiang and Jiangsu Provinces. Ornamental. Hardy to −15°C.

Phyllostachys praecox f. *notata* S. Y. Chen et C. Y. Yao, f. nov.

Grooves on culms yellow-striate.

Distributed in Anji Prefecture of Zhejiang Province. Ornamental.

Phyllostachys praecox f. *prevernalis* S. Y. Chen et C. Y. Yao

The middle of internodes somewhat smaller in diameter than both ends.

Distributed in Anji Prefecture of Zhejiang Province. Ornamental.

16. *Phyllostachys pubescens* Mazel ex H. de Lehaie

Giant bamboos. Culms over 20 m tall, 6 to 15

cm in diameter. Internodes on midculm reaching 40 cm in diameter, (those near culm base shorter). Sheath scars prominent. Young culms dense-pubescent, pruinose. Old culms glabrate, pruinose zone under nodes becoming blackish. Leaves 2 to 3 on twigs. Leaf blades small, lanceolate, 4 to 11 cm long, 0.5 to 1.2 cm wide, of thin texture.

Distributed in the south of Mount Qinling, Valley of Hanshui River to the south of Yangtze River. Ornamental. Hardy to −15°C.

Phyllostachys pubescens f. *gracilis* W. Y. Hsung

Culms smaller than the sp., 7 to 8 m tall, 3 to 4 cm in diameter. Culm wall thicker than the sp.

Distributed in Yixing Prefecture of Jiangsu Province. Ornamental.

Phyllostachys pubescens f. *huamozhu* Wen

Internodes of culms and primary branches yellow, with green-longitudinal-stripes.

Distributed in Mount Mogan-shan of Zhejiang Province. Ornamental.

Phyllostachys pubescens var. *heterocycle* (Carr) H. de Lehaie

Smaller than the sp. Internodes on lower culms shortened and swollen, connected with each other in rhombus-shape pattern.

Cultivated plant, highly ornamental.

17. *Phyllostachys kwangsiensis* W. Y. Hsiung, Q. H. Dai et J. K Liu, sp. nov.

Culms 8 to 16 m tall, 4 to 10 cm long. Internodes on each culm somewhat equal in length, 25 to 35 cm long. Young culms green, densely hairy. Pruinose zones both above and under nodes. Old culms greenish-yellow or yellow. Culm sheath purplish-brown, longer than internode, sparsely dark-brown-maculate. Culm sheaths of lower culms covered with purplish-brown hairs, those of upper culms nearly glabrous. Primary branches 2 on nodes, one branch extra large (sheath blades lorate, 30 cm long), the other small. Leaves 1 to 4 on twigs. Leaf blades lorate-lanceolate, 10 to 15 cm long, 0.8 to 1.5 cm wide, sparse-puberulent, whitish-green beneath.

Distributed in Guangxi Province. Tender.

18. *Phyllostachys circumpilis* C. Y. Yao et S. Y. Chen, sp. nov.

Culms 5 to 7 m tall, 3 to 4.5 cm in diameter. The longest internodes 17 cm long. Young culms dark green, sheath scars prominently puberulent. Old culms grayish-green to grayish-whitish, irregularly pale-orange-yellow-striate and maculate. Culm sheaths pale yellowish-green, markedly-pale-purple-lineate and brown-maculate, densely grayish-glochidiate, ciliate on margins. Leaves 2 to 3 on twigs. Leaf blades dark green, 7.8 to 12 cm long, 1.8 to 2 cm wide.

Distributed in Zhejiang Province. Hardy to −7°C.

19. *Phyllostachys nigra* (Lodd.) Munro

Culms 3 to 6 m tall, 2 to 4 cm in diameter. Young culms green, pubescent, pruinose but turning blackish after one year, then glabrate. Sheath auricles falciform. Sheath ligules long and protruding at apex. Leaves 2 to 3 on twigs. Leaf blades of thin texture, slender-lanceolate, 4 to 10 cm long, 1 to 1.5 cm wide.

Distributed in the Valley of Yangtze River. Hardy to −20°C. Highly ornamental. Found in Beijing Gardens.

Phyllostachys nigra var. *henonis* (Mitf.) Stapf ex Rendle

Culms 7 to 18 m tall, green to grayish-green. Culm wall thicker than the sp.

Distributed in Central China, southwestern China, Hunan, and Gansu Provinces. Hardy to −20°C, tolerates poor soil. Vertical distribution up to 1200 m high.

20. *Phyllostachys aureosulcata* McClure

Culms 3 to 6 m tall; 3 to 4 cm in diameter. Young culms densely pubescent. Grooves on culm internodes yellow. Culm sheaths pale yellow; whitish-green-striate; sparsely-purplish-brown-maculate or none. Sheath auricles prominent.

Cultivated plant, ornamental. Hardy to −20°C. Found in Beijing Gardens.

Phyllostachys aureosulcata f. *spectabilis* C. D. Chu et C. S. Chao

Culms golden yellow, longitudinal grooves green.

Distributed in Mount Yuntaishan of Jiangsu Province. Highly ornamental.

Phyllostachys aureosulcata f. *alata* Wen

Internodes green. Culm sheath sometimes maculate.

21. *Phyllostachys decora* McClure

Culms 8 to 9 m tall, 4 to 5 cm in diameter. The longest internodes on midculm 36 cm long. Young culms bright green, white-glochidiate, nearly not pruinose; sparsely-hairy and an indistinct pruinose zone under nodes. Old culms green or yellowish-green. Nodes slightly swollen. Leaves 2 to 3 (rarely 1) on twigs. Leaf blades lorate-lanceolate to lanceolate, 5 to 12 cm long, 1 to 2 cm wide, pale green and pubescent on the base beneath.

Distributed in the south Yellow River Valley to the valley of Yangtze River. Can grow on sandy soil. Hardy to −15°C.

22. *Phyllostachys viridi-glaucescens* (Carr.) A. et C. Riv.

Culms 6 to 10 m tall, 2 to 5 cm in diameter, green. Young culms dark green, thick-pruinose. Not prominently longitudinally-ribbed on internodes. Culm sheaths pale-brownish-purple, prominently setose and maculate (blotches larger on the top). Sheath auricles purplish-brown or pale-purplish-green; cilia pale green, long and dense. Sheath blades lorate, plicate on the upper part, reflexed. Leaves 2 to 5 on twigs. Leaf blades 4 to 14 cm long, 0.7 to 2 cm wide, usually glabrous.

Distributed in Jiangsu and Zhejiang Provinces. Hardy to −12°C.

23. *Phyllostachys elegans* McClure

Culms 10 m tall, 5.4 cm in diameter, glabrous. Internodes on midculm to 30 cm long (shorter on lower culm); narrow-striate; pruinose at earlier stage. Nodal ridges and sheath scars prominently swollen. Sheath blades lorate-lanceolate, longitudinal-grooved, usually rugose and recurved. Culm sheaths glabrous or sparsely setose on abaxial surface, densely brown-maculate. Leaf blades lanceolate, 8 to 15 cm long, 12 to 20 mm wide.

Distributed in Hannan Island and other areas of Guangdong Province. Tender.

24. *Phyllostachys bambusoides* Sieb. et Zucc.

Culms 10 to 20 m tall, 8 to 10 cm in diameter. Young culms green, usually not pruinose. Old culms dark green. Nodal ridges slightly swollen. Culm sheaths yellowish-brown, with purplish-brown spots and blotches; sparsely setose, later glabrate. Sheath auricles 1 or 2, small, sparsely fimbriate-ciliate. Sheath auricles and ligules yellowish-green or purplish. Leaves 4 to 6 at first, then only 2 to 3 after the others are shed from twigs. Leaf blades lorate-lanceolate, 8 to 20 cm long, 1.3 to 3 cm wide, dark green above, whitish-green beneath, hairy at base.

Distributed from the south of the Yellow River Valley to the south of Yangtze River Valley. Hardy to −18°C.

Phyllostachys bambusoides f. *tanaka* Makino ex Tsuboi

Culm purplish-brown-maculate, branches also maculate.

Distributed in the same area as the sp. Ornamental.

25. *Phyllostachys dulcis* McClure

Culms 7 m tall, 4 to 5 cm. in diameter. The longest internodes on midculms 24 m long. Young culms green, glabrous, pruinose zone under nodes. Nodal ridges slightly swollen. Culm sheaths pale yellow, sparsely pale-brown-maculate. Sheath auricles ovate to falciform, with a few long cilia. Sheath ligules two times broader than the sheath blades, short-ciliate. Sheath blades broad-lorate, pale-purplish-red. Leaves 2 to 4 on twigs. Leaf blades broad-lorate-lanceolate, 10 to 16 cm long, 1.5 to 2.5 cm wide, densely pubescent beneath, especially dense on the base.

Distributed in Jiangsu and Zhejiang Provinces. Bamboo shoots delicious. Hardy to −12°C.

26. *Phyllostachys prominens* W. Y. Xiong, sp. nov.

Culms 10 m tall, 7 cm in diameter. Internodes on midculm 22 cm long. Except several internodes near the culm base, all the others somewhat of equal length. Young culms dark green, later turning green, not pruinose. Culm sheaths pale-brownish-yellow, or slightly pale-reddish; densely blackish-brown-maculate, especially dense at top; spots aggregated into large blotches on the lower part. Sheath ligules purplish-black, wavy at apex. Sheath blade lorate-lanceolate, orange-red or green. Leaves 2 to 3 on twigs. Leaf blades lorate-lanceolate, 8.5 to 18 cm long, 1.3 to 2.2 cm wide, hairy on the base beneath.

Distributed in Zhejiang Province. Hardy to −7°C.

27. *Phyllostachys yunhoensis* S. Y. Chen et C. Y. Yao, sp. nov.

Culms 5 to 6 m tall, 3 to 4 cm in diameter. Internodes on midculm 13 to 14 cm long. Young culms green; old culms paler, grayish-farinose. Narrow farinose-zone under sheath scars. Culm sheaths dark green to brownish-yellow, densely dark-reddish-brown-maculate, smooth, glabrous, slightly pruinose. Sheath auricles deciduous, green, falciform or ovate, long-purple-ciliate. Sheath blades lorate, reflexed. Leaves 2 to 3 on twigs. Leaf blades lanceolate, 9.5 to 14 cm long, 1.6 to 1.9 cm wide; lateral veins 4 to 5 pairs.

Distributed in Zhejiang Province. Hardy to −7°C.

28. *Phyllostachys platyglossa* C. P. Wang et Z. H. Yu, sp. nov.

Culms 8 m tall, 3.5 cm in diameter. The longest internodes 38 cm long, dark-purplish-green. Culm sheaths slightly farinose, pale-reddish-brown with shade of pale green, sparsely brown-maculate, sparsely hispidulous. Margins of sheath purple, glabrous. Sheath auricles ovate to falciform, purple, long-purple-ciliate. Sheath ligules short and broad, truncate to arcuate at apex, purple. Sheath blades

Phyllostachys bambusoides f. *tanaka* Makino ex Tsuboi

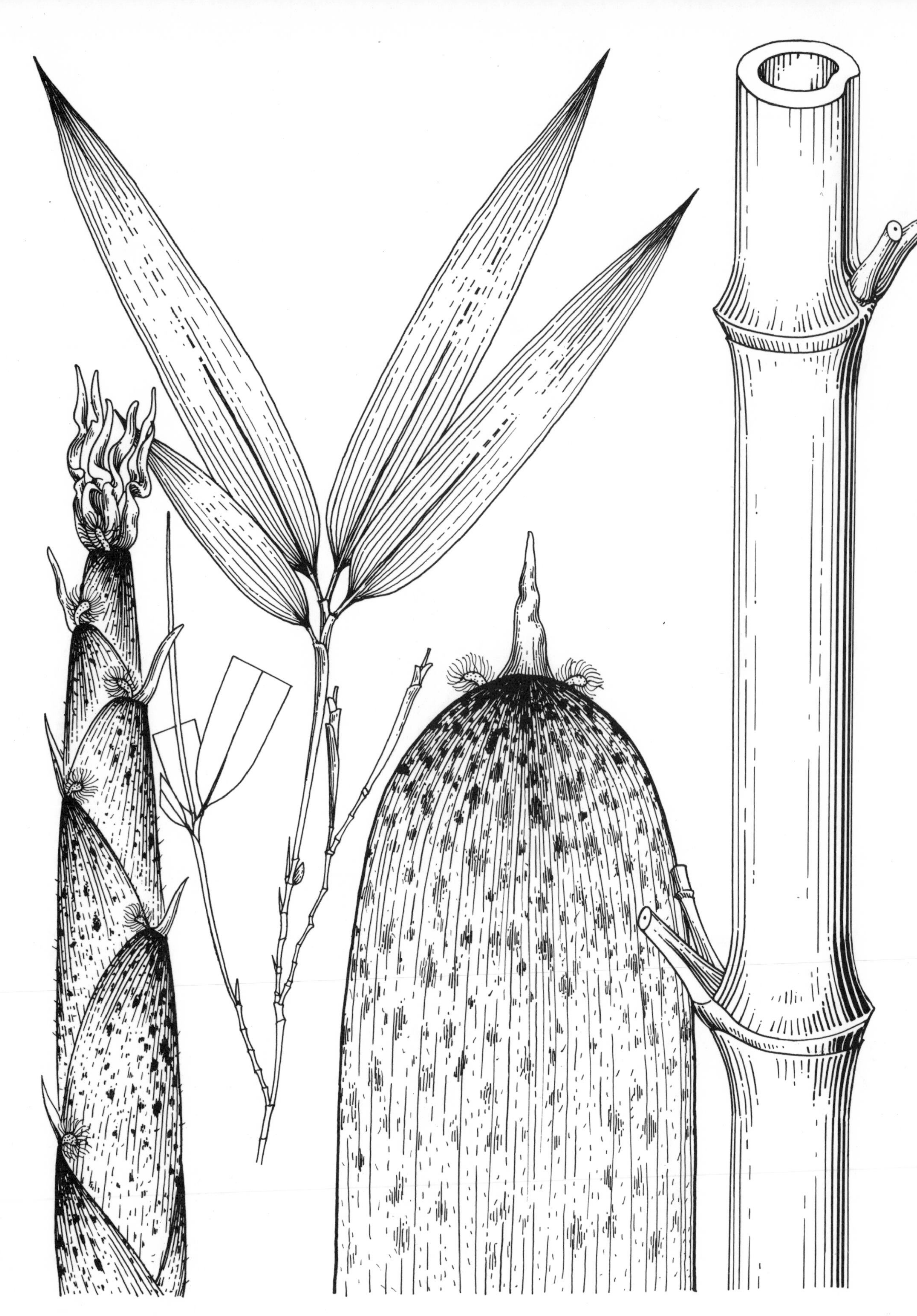

Phyllostachys prominens W.Y. Xiong

lorate, greenish-purple to green, rugose, reflexed. Leaves 2 on twigs. Leaf blades 7 to 14 cm long, 1.2 to 2.2 cm wide; lateral veins 5 to 7 pairs.

Distributed in Anji Prefecture of Zhejiang Province. Hardy to −7°C.

29. *Phyllostachys nidularia* Munro

Culms 8 m tall, 4 cm in diameter; crowded; crown narrow, conical. The longest internodes on midculm 40 cm long. Young culms green, sometimes purple-striate. Old culms also green. Nodal ridges and sheath scars equally slightly swollen. Culm sheaths somewhat hairy, denser on the base. Sheath auricles prominently large. Leaves 1 (rarely 2) on twigs. Leaf blades oblong-lanceolate to lanceolate, 7 to 13 cm long, 1.3 to 2 cm wide, slightly drooping.

Distributed to the south of Mount Qinling of Shaanxi Province and the valley of Yangtze River. Hardy to −15°C.

Phyllostachys nidularia f. *farcta* H. R. Zhao et A. T. Liu, f. nov.

Culms solid or nearly solid.

Distributed in Guangdong Province. Tender.

30. *Phyllostachys guizhouensis* C. S. Chao et J. Q. Zhang, sp. nov.

Culms 16 m tall, 8 cm in diameter. Young culms green, sparsely short-setose, scabrid. Old culms grayish-green, pruinose under nodes. Internodes 30 to 41 cm long. Nodal ridges on lower culm not swollen, but those on upper culm swollen. Sheath blades narrow-lorate, upright or spreading, purplish-brown, green striate. Leaf sheaths glabrous, oral setae sparse and upright, fugacious. Leaf blades lanceolate, 8 to 11 cm long, 1.1 to 1.6 cm wide.

Distributed at an altitude of 1440 m on streamsides in Biji Prefecture of Guizhou Province. Tolerates a minimum temperature of −10°C.

31. *Phyllostachys robustiramea* S. Y. Chen et C. Y. Yao, sp. nov.

Culm erect, 2 to 3 m tall, 2 to 3 cm in diameter. The longest internodes 19 cm long. (Extra large culms 10 m tall, 6 cm in diameter.) Usually one longitudinal groove on the side opposite to the branching side of internodes, making the internodes compressed and ribbed, also with small longitudinal furrows. Young culms purplish-green, pruinose, glabrous, turning to pale green later, farinose zone under sheath scars. Nodal ridges very swollen while sheath scars only slightly swollen. Culm sheaths of thin texture, greenish-purple, with creamy-purple-radiating stripes at apex, sparsely puberulent; pruinose on the base. Sheath ligules pale green, truncate or slightly arcuate at apex. Branching from the 3rd to 7th nodes up from culm base. The main branch much larger than the others. Leaves 3 on twigs. Leaf auricles prominent, ciliate. Leaf blades 6.2 to 11.8 cm long, 1.1 to 1.4 cm wide.

Distributed in Anji Prefecture of Zhejiang Province. Hardy to −7°C.

32. *Phyllostachys stimulosa* H. R. Zhao et A. T. Liu, sp. nov.

Culms 8 m tall, 3.4 cm in diameter. Internodes 20 cm long, but the longest reach 32 cm; smooth to scabrid. Nodal ridges as swollen as sheath scars. The margins of culm sheaths yellowish-brown. The base of sheath blades decurrent into broad-ovate auricles. Sheath auricles purple, ciliate. Sheath ligules short, arcuate at apex, short-ciliate. Sheat blades greenish-purple, erect. Leaves 1 to 3 on twigs. Leaf blades of thick texture, dark green, but grayish-white beneath; 6 to 15.5 cm long, 1 to 2 cm wide.

33. *Phyllostachys heteroclada* Oliver

[*Ph. congesta* Rendle]

Culms 1 to 4 m tall, 1 to 2.5 cm in diameter on lower culm; upper culm slender; green, turning to yellowish-green in the second year; glabrous. Young culms sparse-pruinose, glabrate later. Nodal ridges slightly more swollen than sheath scars. Intranodes 5 mm wide. Branches spreading. Culm sheaths green, glabrous, except ciliate, cilia white or pale-yellow. Sheath ligules short, yellowish-green or purplish-red. Sheath blades triangular to triangular-lanceolate, green, erect, appressed to the culm. Leaves 1 to 3 on twigs. Leaf sheath glabrous, but oral cilia long, fugacious. Leaf blades oblong-lanceolate, 7 to 10 cm long, 1.3 to 1.6 cm wide, of thin texture, hairy on the base beneath, with transverse veinlets.

Distributed to the south of the Yangtze River. Hardy to −10°C.

Phyllostachys heteroclada f. *solida* (S. L. Chen) C. P. Wang et Z. H. Yu, f. nov.

Culms solid or nearly solid.

Distributed in Zhejiang, Jiangsu, Anhui, and Hunan Provinces.

34. *Phyllostachys bissetii* McClure

Similar to *Ph. heteroclada* in appearance, but upper part of the internodes of young culms sparsely-erect-hairy; the base of sheath blades distinctly narrower than sheath ligules.

Distributed in Zhejiang and Sichuan Provinces. Hardy to −5°C.

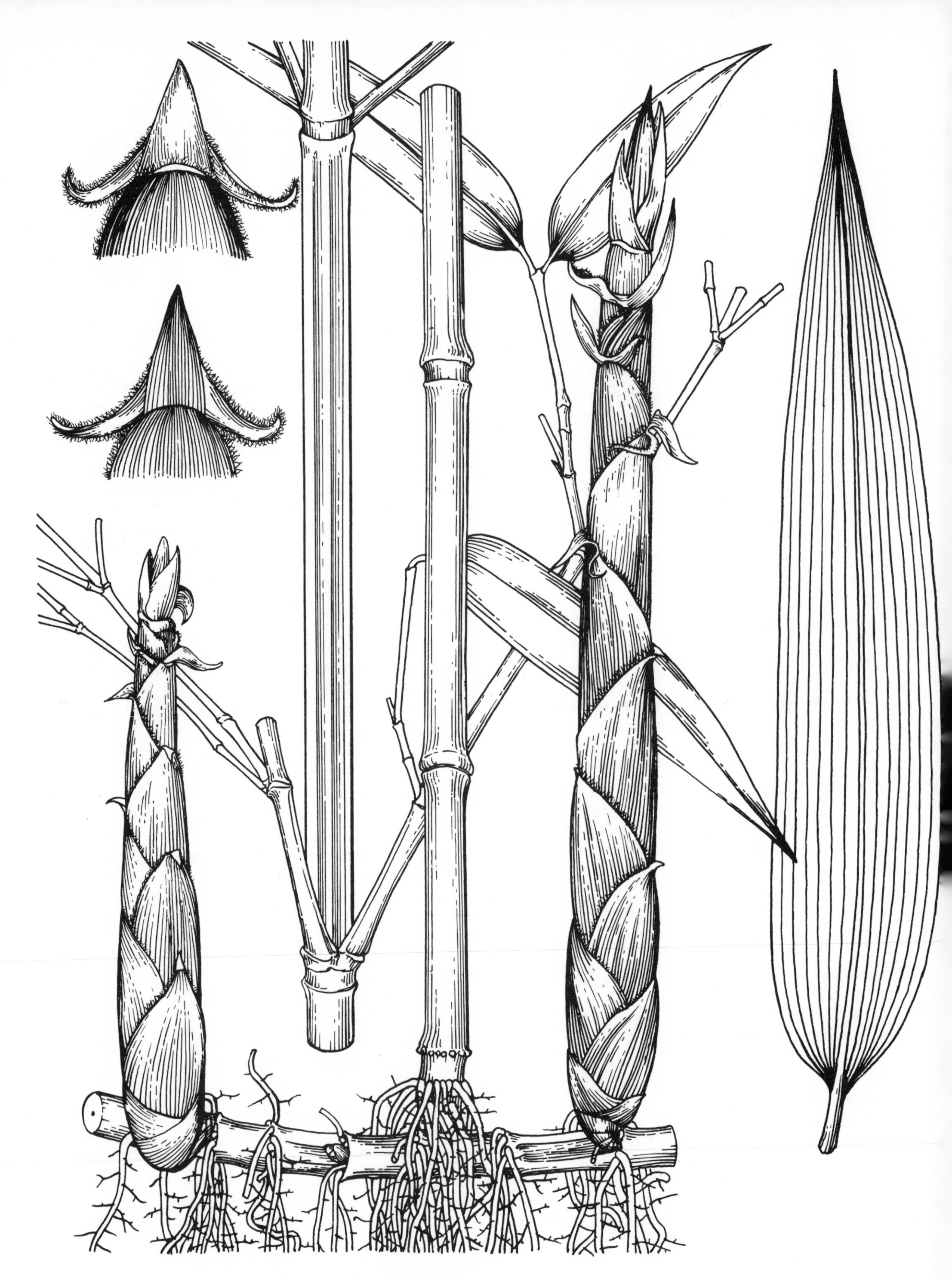

Phyllostachys nidularia Munro

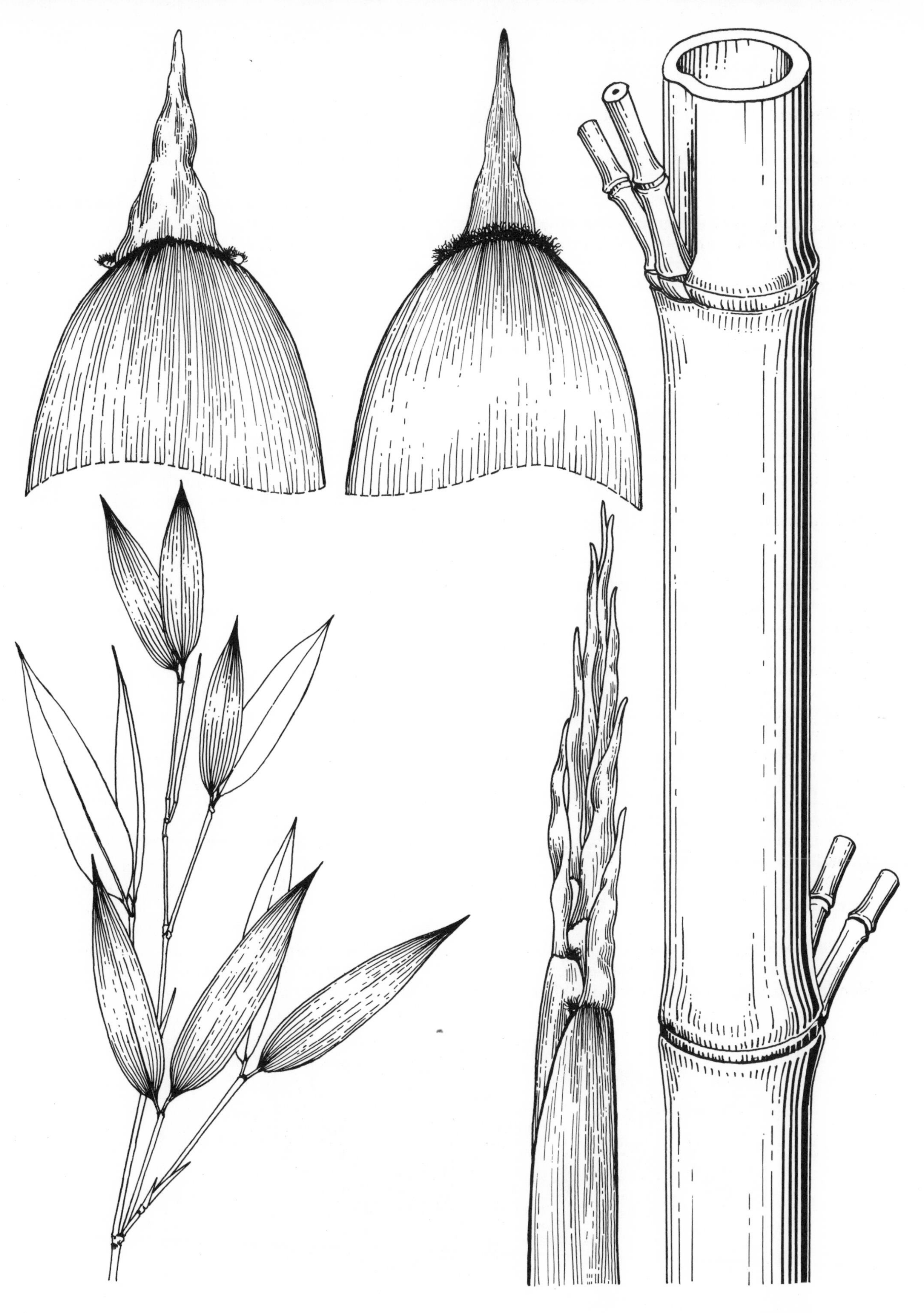

Phyllostachys parvifolia C.D. Chu et H.Y. Chou

35. *Phyllostachys parvifolia* C. D. Chu et H. Y. Chou

Culms 8 m tall, 5 cm in diameter. The longest internodes at midculm 24 cm long. Young culms green; purple-lineate; densely pruinose, especially under nodes. Old culm gray. Nodal ridges slightly swollen. Sheath scars on the lower culms more swollen than nodal ridges. Culm sheaths pale brown or pale purplish-red; pale-yellowish-brown or whitish-yellow-lineate; not maculate, glabrous but white-ciliate. Sheath ligules pale purplish-red, as broad as the base of sheath blades, serrulate, densely short-ciliate. Leaves 2 (rarely 1) on twigs. Leaf blades small, lanceolate to lorate-lanceolate, 3.5 to 6.2 cm long, 0.7 to 1.2 cm wide, hairy on the base beneath.

Distributed in Anji Prefecture of Zhejiang Province. Hardy to −7°C.

36. *Phyllostachys rivalis* H. R. Zhao et A. T. Liu, sp. nov.

Culms over 4 m tall, 1.5 to 2 cm in diameter. Internodes on midculm 20 cm long. Old culms brownish-yellow, somewhat purplish. Young culms brownish-purple or yellowish-green, indistinctly purple-striate, pruinose, pubescent. Nodal ridges swollen, purple when young. Sheath scars sparsely white-ciliate when young. Culm sheaths thick-papery, green when young, turning to brownish-purple later, brown-small-maculate and indistinct-purple-striate on abaxial surface, pale-brown-ciliate on rim. Leaves (2) 3 to 5 (7) on twigs. Leaf blades small, somewhat hard, oblong-lanceolate, 4.6 to 7.2 cm long, 0.6 to 1.1 cm wide, glabrous above, densely white-pubescent beneath. Lateral veins 3 to 4 pairs.

Distributed in Guangdong and Fujian Provinces. Tender.

37. *Phyllostachys rubicunda* Wen, sp. nov.

Culms 3 m tall, 7 mm in diameter. Internodes slender, curving; nodes very swollen. Culm sheaths glabrous, bright red at first, turning to straw color later. Leaves 2 to 3 on twigs. Leaf sheaths bright red at first, straw color later, long-ciliate at apex. Leaf ligules very short, ciliate. Leaf blades lanceolate, 5 to 6 cm long, 12 to 14 cm wide, lateral veins 5 to 6 pairs.

Distributed in Anji Prefecture and Hanzhou of Zhejiang Province. Hardy to −7°C.

38. *Phyllostachys atrovaginata* C. S. Chao et H. Y Chou, sp. nov.

Culms 7 to 8 m tall, 3 to 5 cm in diameter. Branches short, crown narrow, conical. Internodes on midculm 29 to 31 cm long. Young culms green, glabrous, indistinctly pruinose. Nodal ridges and sheath scars moderately swollen. Leaves 2 to 3 on twigs. Leaf auricles and oral setae obsolescent. Leaf ligules densely pubescent on the back, sometime with long thick hairs on the base. Leaf blades 5.5 to 13 cm long, 0.9 to 1.6 cm wide, lorate-lanceolate; the lowest leaf blade on twigs shorter, lanceolate; lateral veins 6 pairs, hairy near the base beneath and on the stalk.

Distributed in Zhejiang Province. Hardy to −7°C.

39. *Phyllostachys concava* Z. H. Yu et C. P. Wang, sp. nov.

Culms 5.7 to 6.2 m tall, 2.7 to 4.5 cm in diameter. The longest internodes 26 to 29.5 cm long. Young culms dark green, somewhat purplish, glabrous, not pruinose or slightly pruinose, but pruinose zone distinct. Old culms smooth, thick-pruinose. Nodal ridges a little more swollen than or as swollen as sheath scars. Branching from the 10th or 11th node up from the culm base. Leaves 3 to 4 on twigs. Leaf blades of thick-leathery texture, 3.6 to 12.5 cm long, 0.6 to 2.2 cm wide, lanceolate. Leaf sheath glabrous or ciliate only, leaf auricles absent but oral setae remain, leaf ligules not protruding.

Distributed in Anji Prefecture of Zhejiang Province. Hardy to −7°C.

XIV. *Chimonobambusa* Makino

Tall or shrub-like bamboos. Rhizomes monopodial. Culms scattered, erect, square-columnated or cylindrical, usually compressed or grooved on branching side. Mass of spine-like, tuberculate-based aerial roots girdle the lowest few nodes. Primary branches 3 on nodes. Culm sheaths persistent, chartaceous. Sheath auricles usually none. Sheath ligules membranous, whole. Sheath blades small, erect, triangular or circular. No distinct suture between the base of sheath blade and sheath proper. Leaf blades scabrous or a little smoother only on one side.

KEY to the Spp. of *Chimonobambusa* Native to China

1a. Culms setulose when young, becoming muricatus after shedding of bristles as verruciform-base remain. Leaf blades papery. 2
1b. Culms glabrous and smooth. Leaf blades of firm texture. 5
2a. Culms sheaths nearly coriaceous. 3
2b. Culms sheaths papery or thin-coriaceous. 4
3a. Culm internodes square-columnated, blunt on edges. Culms sheaths purple-maculate on abaxial surface, glabrous, ciliate only. Leaves 2 to 5 on twigs. .... (1) *C. quadrangularis*
3b. Culm internodes cylindrical. Culm sheaths with sparse-appressed-bristles on abaxial surface, but especially dense on base; densely ciliate on rim, not maculate. Leaf single on twigs. .... (2) *C. hejiangensis*
4a. Culms internodes square-columnated, blunt on edges. Culm sheaths papery, longer than internodes, sparsely setose on abaxial surface, densely ciliate, not maculate. .... (3) *C. hirtinoda*
4b. Culm internodes cylindrical, but those between branching nodes 4-ridged and 3-canaliculate. Culms sheaths thin-coriaceous, shorter than internodes, grayish-white setulose on abaxial surface, densely grayish-brown-ciliate, not maculate. .... (4) *C. metuoensis*
5a. Culm internodes cylindrical or slightly square-columnated. Culm sheaths thick-papery, brown-maculate on abaxial surface, glabrous or white-pubescent only at base. Leaves 2 to 3 on twigs. .... (5) *C. utilis*
5b. Culm internodes slightly square-columnated. Culm sheaths papery, not maculate on the abaxial surface, glabrous except densely ciliate. Leaves 1 to 2 on twigs. .... (6) *C. angustifolia*

1. *Chimonobambusa quadrangularis* (Fenzi) Makino

Culms 3 to 8 m tall, 1 to 4 cm in diameter. Internodes 8 to 22 cm long. Culm ridges prominently swollen. Sheath scars setulose. Primary branches 3 on nodes at early stage, becoming multi-branched and clustered after accessory buds developed. Branches smooth, branch nodes prominently swollen. Twigs 1 to many clustered on branch nodes, nearly solid, twig nodes prominently swollen. Leaves 2 to 5 on twigs. Leaf sheaths coriaceous, glabrous, oral setae remain. Leaf ligules short, truncate, fine-ciliate, setulose on abaxial surface. Leaf blades thin-papery, narrow-lanceolate, 8 to 29 cm long, 10 to 27 mm wide; acute at apex, narrowed at base to stalks; glabrous above, pubescent beneath, glabrate later. Lateral veins 4 to 7 pairs, transverse veinlets distinct.

Distributed on the south slopes of Mount Qinling of Shaanxi Province, eastern, southern, and southwestern China. Hardy to −15°C and highly ornamental.

2. *Chimonobambusa hejiangensis* C. D. Chu et C. S. Chao, sp. nov.

Culms 5 to 7 m tall, 2 to 3 cm in diameter, green, white-pubescent and sparsely hispid when young, glabrate afterwards, nodes purple. Internodes 20 cm long. Leaf single on twigs. Leaf sheaths adnate. Leaf auricles and oral setae obscure. Leaf blades papery, linear-lanceolate, 11 to 20 cm long, 1.5 to 2 cm wide, glabrous. Lateral veins 4 to 6 pairs.

Distributed in Sichuan Province at an altitude of about 1000 m. Hardy to −7°C.

3. *Chimonobambusa hirtinoda* C. S. Chao et K. M. Lan, sp. nov.

Culms 5 m tall, 2.5 cm in diameter. Internodes 12 to 16 cm long. Sheath scars densely brown-

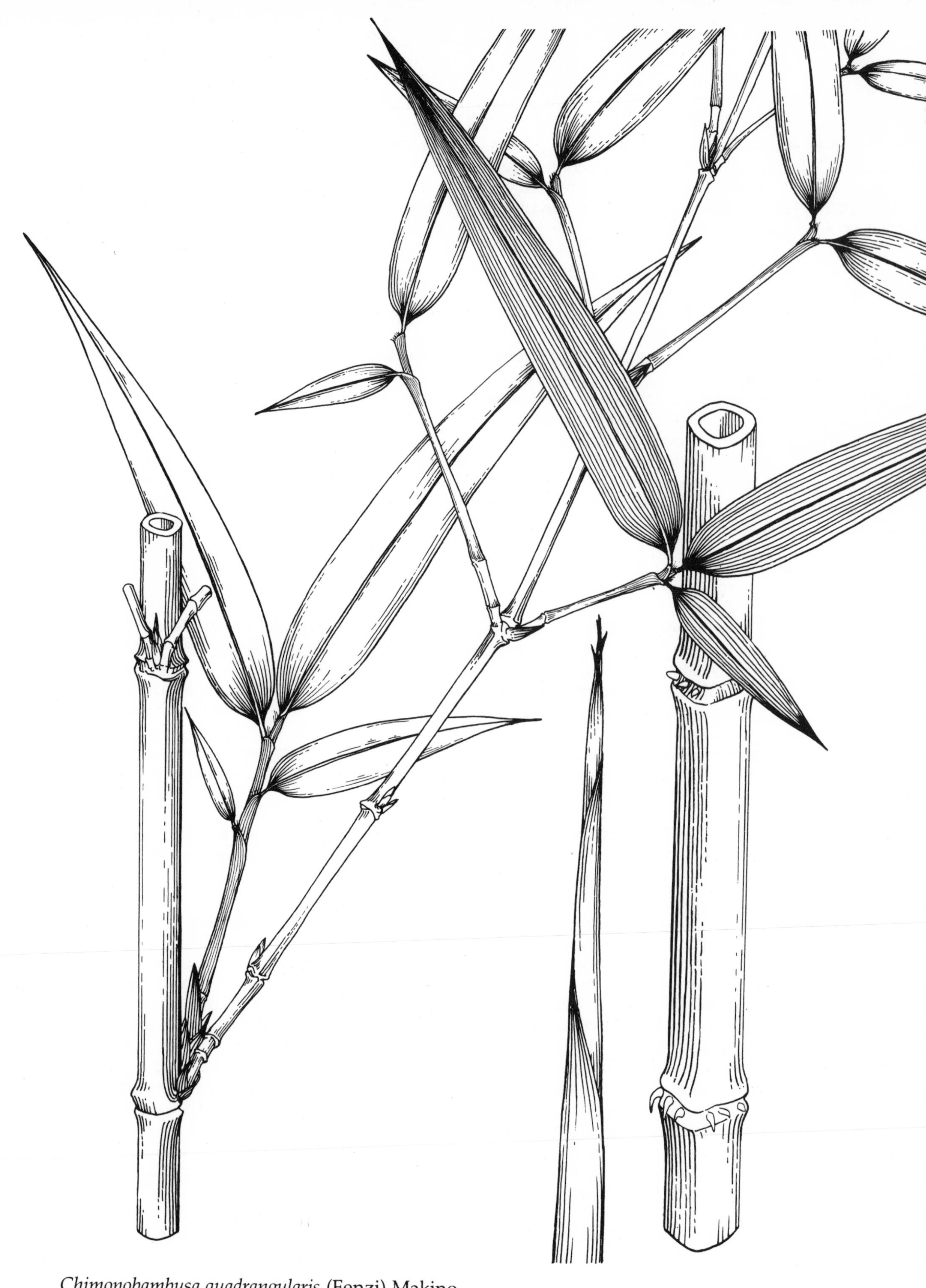

Chimonobambusa quadrangularis (Fenzi) Makino

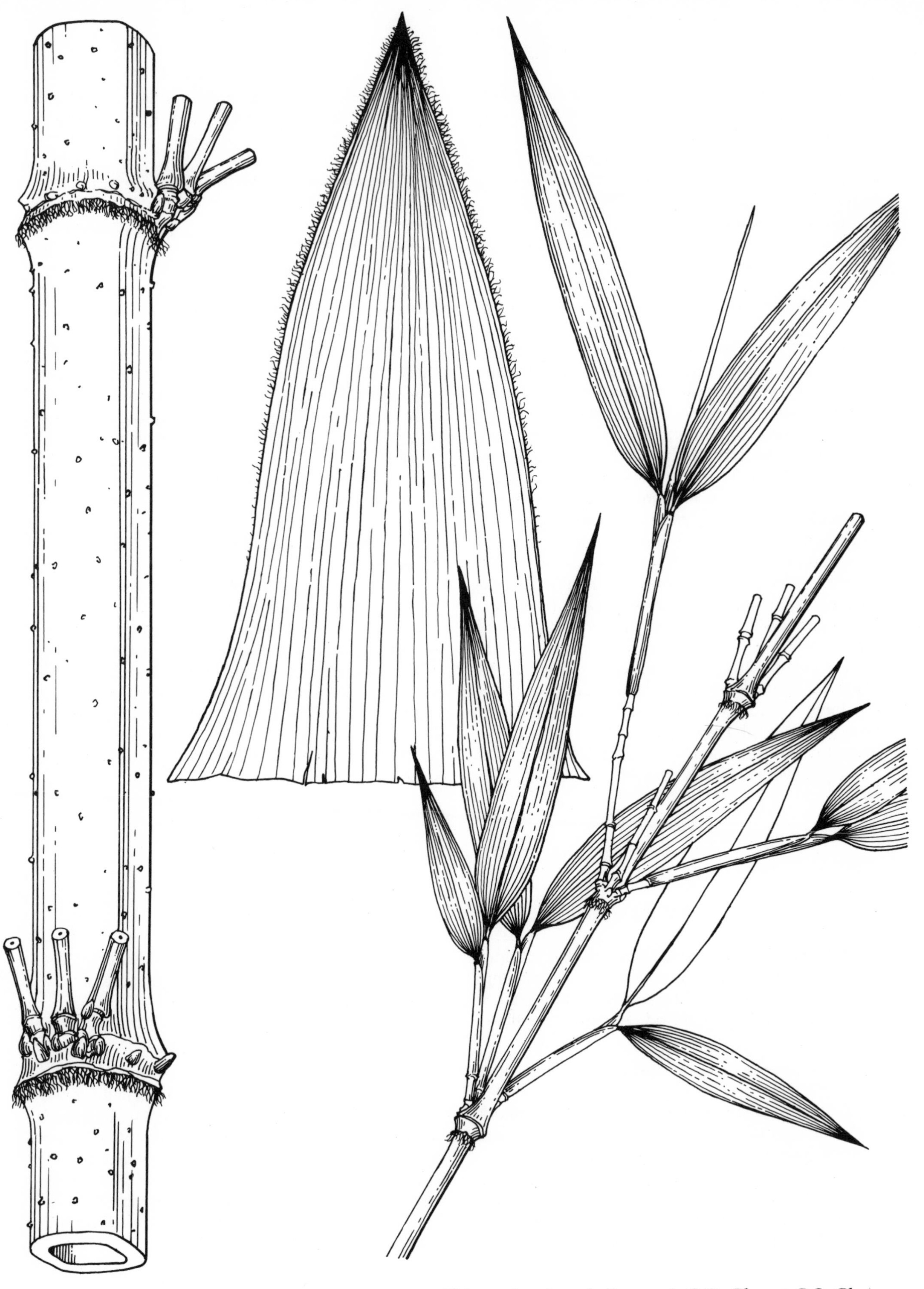

Chimonobambusa hejiangensis C.D. Chu et C.S. Chao

macrosetose, persistent. Vein meshes on sheaths distinct. Leaves 2 to 3 on twigs. Leaf sheath glabrous. Oral setae long, upright, sparse. Leaf blades lorate-lanceolate to lanceolate, 9 to 11 cm long, 1 to 1.8 cm wide; acuminate at apex; glabrous. Lateral veins 4 to 5 pairs.

Distributed at an altitude of 1100 m on Mount Doupengshan in Douyun Prefecture, Guizhou Province. Can tolerate a few degrees of frost.

4. *Chimonobambusa metuoensis* Hsueh et Yi, sp. nov.

Culms 5 to 7 m tall, 1 to 2.5 cm in diameter. Internodes 20 to 28 cm long, green. Sheath scars swollen, densely brown-setulose when young, (setulose zone under nodes,) glabrate afterwards. Nodal ridges prominently swollen, glabrous. Intranodes densely yellowish-brown-pubescent when young. Branching usually beginning at the 15th node from the base. Primary branches at first 3 on nodes, becoming multi-branched after the development of accessory buds. Primary branches 30 to 60 cm long, with 8 to 12 nodes, glabrous, branch nodes swollen. Leaves 2 to 3 on twigs. Leaf sheaths glabrous, longitudinal-lineate, longitudinal-ridged, distinctly on upper part, which is dark purple when young. Leaf auricles obscure. Leaf ligules slightly convex or truncate, deep purple, glabrous. Leaf blades lanceolate, papery, 12 to 33 cm long, 1.5 to 4 cm wide; acuminate at apex, broad-cuneate at base; green above, pale green beneath; glabrous; serrulate. Lateral veins 5 to 8 pairs, transverse veinlets distinct.

Distributed at an altitude of 1900 to 2200 m in Xizang (Tibet) Autonomous Region. Hardy to −7°C.

5. *Chimonobambusa utilis* (Keng) Keng f.

Culms 3 to 10 m tall, 1 to 3.5 cm in diameter. Internodes 9 to 30 cm long. Nodes prominently swollen. Primary branches 3 on nodes. Branch nodes prominently swollen. Leaves 2 to 3 on twigs. Leaf sheaths glabrous, oral setae none. Leaf ligules truncate or arcuate, slightly hairy on abaxial surface. Leaf blades firm, smooth, 5 to 14 cm long, 2 to 23 mm wide; acuminate at apex, narrowed at base to stalk; dark green above, whitish-green beneath; one side of leaf rim setulose and scabrous, the other side slightly scabrous or nearly smooth near base. Lateral veins 5 to 7 pairs, transverse veinlets distinct.

Distributed at an altitude of 1000 m in Sichuan Province. Hardy to −5°C.

6. *Chimonobambusa angustifolia* C. D. Chu et C. S. Chao, sp. nov.

Culms 2 m tall, 1 cm in diameter, purplish-green when young, dark green afterwards. Internodes 8 to 15 cm. Nodes purple, slightly swollen. Primary branches 3 on nodes. Branch nodes swollen, geniculate. Leaves 1 to 2 (rarely 3) on twigs. Leaf sheaths ciliate, oral setae a few. Leaf blades linear to linear-lanceolate, 8 to 17 cm long, 7 to 12 mm wide, glabrous. Lateral veins 3 to 4 pairs.

Distributed at an altitude of 1100 to 1200 m in Guangxi Province. Tender.

XV. *Acidosasa* C. D. Chu et C. S. Chao, gen. nov.

Tall bamboo, rhizomes monopodial. Culms erect, cylindrical, slightly grooved on the branching side. Primary branches 3 on midculm nodes, 5 on upper-culm nodes. Nodal ridges slightly swollen. Culm sheaths deciduous. Sheath blades small. Leaf blades variable in size, usually large. Lateral veins many, reticulate veins clearly visible.

Acidosasa chinensis C. D. Chu et C. S. Chao, sp. nov.

Culms 8 m tall, 3 to 5 cm in diameter, internodes on midculm about 20 cm long. Culms green, densely hispidulous at earlier stage, then grabrate; narrow-striate. Nodal ridges and sheath scars slightly swollen. The apex of bamboo shoots flat. Culm sheaths crisp, brownish-red, hispidulous (fugacious), sparsely maculate, ciliate, acuminate at apex; reticulate veins clearly visible. Sheath auricles and oral setae absent. Sheath blades small, lanceolate, 1.5 to 4.5 cm long, less than 1 cm wide. Sheath ligules short, arcuate and short-fimbriate-ciliate at apex. Leaves 2 to 5 on twigs. Leaf blades usually large, acuminate at apex, narrowed at base, serrulate, glabrous, lateral veins 6 to 11 pairs.

Distributed in Yangchuan Prefecture of Guangdong Province. Tender.

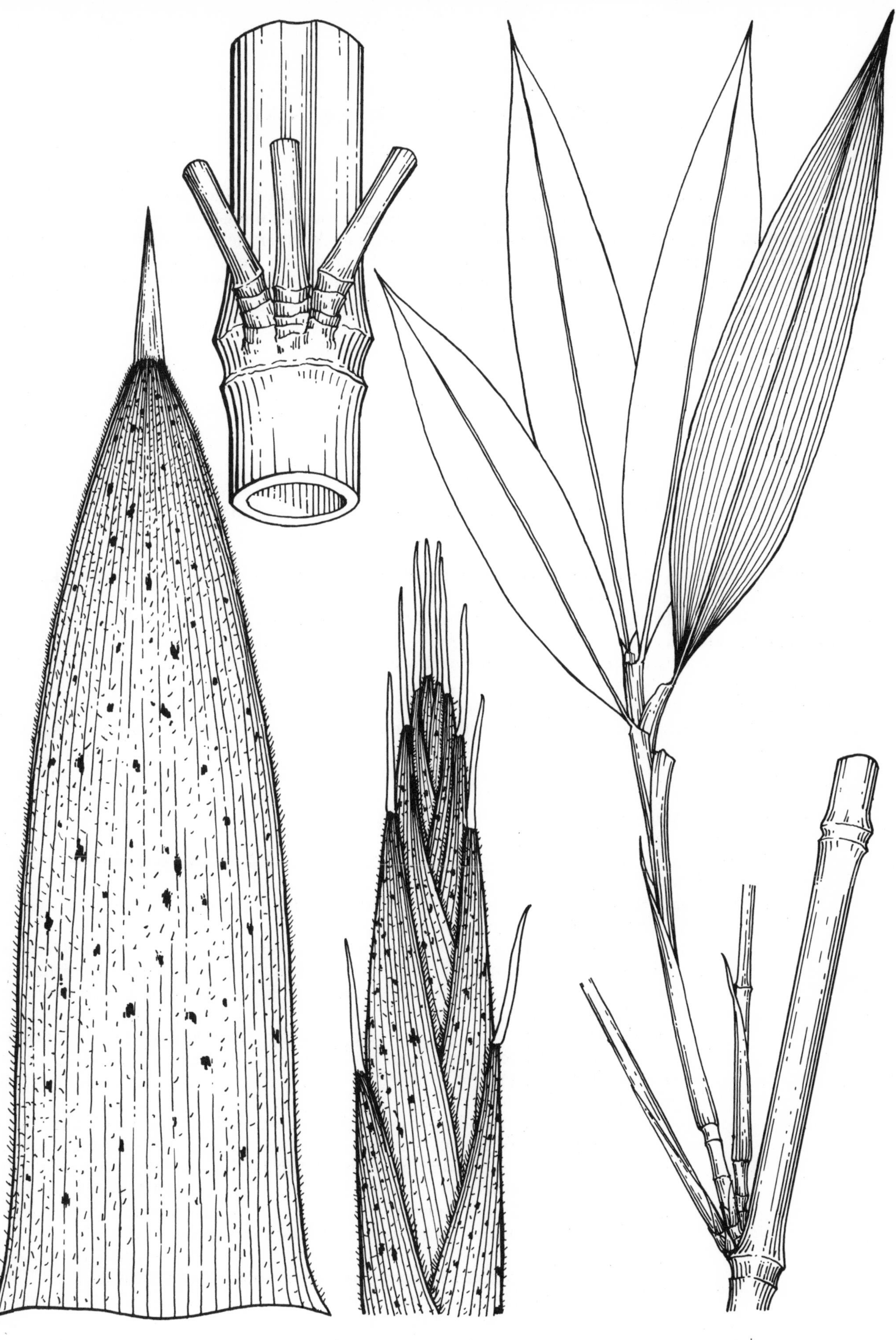

Acidosasa chinensis C.D. Chu et C.S. Chao

XVI. *Sinobambusa* Makino ex Nakai

Tall or shrub-like bamboos, rhizomes amphipodial; culms scattered or in scattered clumps, erect. Internodes long, cylindrical, compressed or grooved from mid-internode downwards on the branching side, lineate. Sheath scars corky and as swollen as nodal ridges, with transverse furrows between them. Primary branches 3 (sometimes 5 to 7) at midculm nodes. All primary branches on the same node of even thickness, spreading; branch nodes swollen. Culm sheaths deciduous, of leathery or thick-papery texture, bristly or glabrescent, but usually with densely strewn bristles on the base; ligules arciform, entire. Sheath blades lanceolate, deciduous; auricles prominent or none. Leaves 3 to 9 on twigs. Leaf blades lanceolate, with transverse veinlets.

KEY to the Spp. of *Sinobambusa* Native to China

1a. Culm sheaths nearly oblong, slightly narrowed at apex, usually without stinging bristles. No pigskin-like pores on or under nodal ridges. 2

1b. Culm sheaths nearly triangular or narrowly triangular, greatly narrowed at apex. With pigskin-like pores under nodal ridges. 13

2a. Culm sheath auricles prominent, ligules yellowish-green. Young culms and culm sheaths not pruinose or scarcely pruinose. 3

2b. Culm sheath auricles obsolescent or absent. 11

3a. Culm sheaths bristly on the abaxial surfaces; auricles slanting in position; oral setae long and thick. 4

3b. Culm sheaths glabrous on the abaxial surfaces, pruinose; auricles erect; oral setae short and thin. . . . . (1) *S. seminuda*

4a. Young culms glabrous; sheath ligules 4 mm long; with short bristles on the base of culm sheaths. 5

4b. Young culm puberulent; sheath ligules shorter; with long fine hairs on the base of culm sheaths. 10

5a. Sheath auricles elliptic to falciform, spreading; or not prominent. Leaf auricles usually obsolescent or absent, seldom spreading. 6

5b. Sheath auricles reniform to elliptic; leaf auricles usually spreading. . . . (2) *S. nephroaurita*

6a. Sheath auricles moderately prominent, spreading; with densely strewn bristles on the base of culm sheaths. Sheath blades not plicate. Leaf blades puberulent or glabrous beneath. 7

6b. Sheath auricles very small, not spreading; slightly puberulent on the base of culm sheaths. Sheath blades plicate. Leaf blades glabrous beneath. . . . . . (3) *S. incana*

7a. Leaf blades puberulent beneath. 8

7b. Leaf blades glabrous beneath. 9

8a. Leaf blades densely pubescent beneath. Sheath ligules 3 to 4 mm long. Sheath blades green. . . . . (4) *S. tootsik*

8b. Leaf blades puberulent beneath. Sheath ligules truncate at apex. Sheath blades purplish. . . . . (4c) *S. tootsik* var. *laeta*

9a. Sheath ligules entire at apex. . . . . (4a) *S. tootsik* var. *tenuifolia*

9b. Sheath ligules serrate or sparsely double serrate at apex. . . . . (4b) *S. tootsik* var. *dentata*

10a. Intranodes smooth. Sheath ligule obtuse at apex, hispidulous. Leaf blades basihirsutus beneath. . . . . (5) *S. intermedia*

10b. Intranodes creased. Sheath ligules truncate at apex. Leaf blades puberulent beneath. . . . . (6) *S. pulchella*

11a. Young culms and culm sheaths thickly pruinose and bristly. Sheath ligules yellowish-green, thickly pruinose. . . . . (7) *S. farinose*

11b. Young culms, culm sheaths and sheath ligules not pruinose or obsolescently pruinose. 12

12a. Sheath auricles none. Long-setose on the base of culm sheaths. Sheath blades and leaf blades puberulent on abaxial surface. . . . . (8) *S. rubroligula*

12b. Sheath auricles very small. Puberulent on the base of culm sheaths. Sheath blades and leaf blades glabrous on both sides. . . . . (9) *S. glabrescens*

13a. Usually without elevated longitudinal lines on internodes, glabrous. Pigskin-like pores under nodes. 14

13b. With elevated longitudinal lines on internodes. No pigskin pores under nodes. 16

14a. Young culm pruinose in whole. Leaf auricles and oral setae none. Leaf blades puberulent beneath. Brown chaetae on the edge of culm sheath. Sheath ligules cuspidate at apex. . . . . (10) *S. gigantea*

14b. Pruinose only under nodes of young culms. Leaf auricles and oral setae prominent. Leaf blades glabrous beneath. No chaetae on the edge of culm sheath. Sheath ligules obtuse at apex. 15

15a. Young culms puberulent. Fine hairs at apex of leaf ligules. Leaf sheaths long-ciliate. Transverse veinlets on both sides of leaf blades. Serrulate on one side only of leaf blades. . . . . (11) *S. puberula*

15b. Young culms glabrous. Leaf ligules glabrous at apex. Leaf sheaths not ciliate. Leaf blades with transverse veinlets beneath only. Serrulate on both sides of leaf blades. . . . . (12) *S. urens*

16a. Bamboo shoots appearing in winter. Culm sheaths hard, ciliate. . . . . (13) *S. scabrida*

16b. Bamboo shoots appearing in summer. Culm sheaths of leathery or thick-papery texture, glabrous on edge. 17

17a. Young culms glabrous. Culm sheaths of thick-leathery texture, pruinose. Oral setae erect, 1 cm long. Sheath blades incurved. . . . . (14) *S. henryi*

17b. Young culms pilose. Culm sheaths of thick papery texture, usually glabrous. No oral setae. Sheath blades erect. 18

18a. Sheath ligules short, decurrent on both sides. Sheath auricles none. Culm sheaths purplish-green. . . . . (15) *S. anaurita*

18b. Sheath ligules cuspidate, not decurrent. Sheath auricles elliptic, ciliate. Culm sheaths green, with longitudinal stripes. . . . . (16) *S. striata*

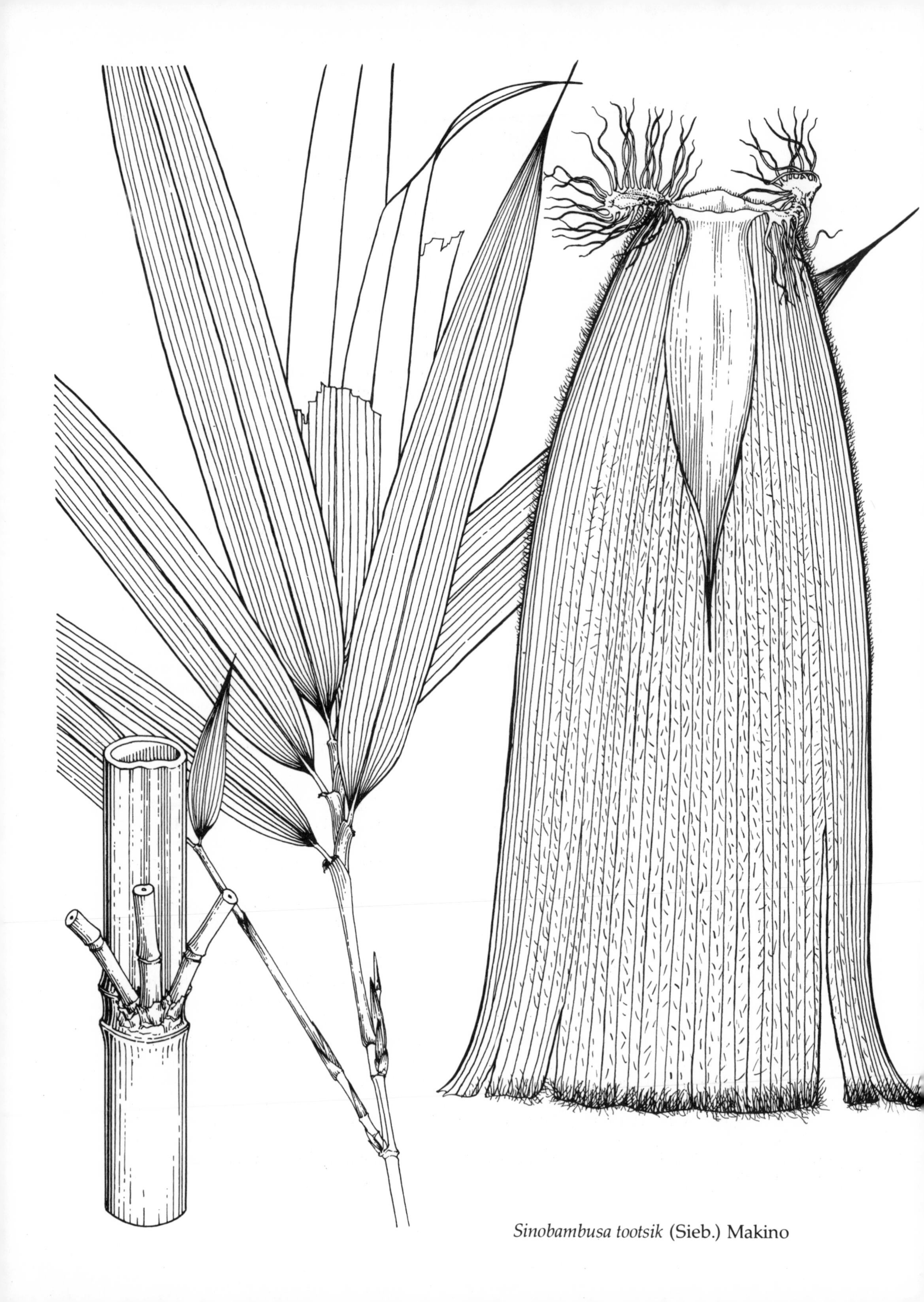

Sinobambusa tootsik (Sieb.) Makino

1. *Sinobambusa seminuda* Wen sp. nov.

Culms 4 m tall, 1.5 cm in diameter; internodes on midculm 40 cm long, half cylindrical or compressed, zigzag. Young culms green, glabrous, lineate, pruinose below nodes. Nodal ridges slightly swollen, liratus; sheath scars corky, puberulent at first, then glabrate. Primary branches 3 at midculm nodes, secondary branches 1 to 3 on primary branch nodes. Leaves 5 to 6 on twigs. Leaf sheaths yellowish-green, with veins elevated longitudinally, also with transverse veinlets, glabrous, except brown-ciliate. Leaf blades lanceolate to broad-lanceolate, 9 to 18 cm long, 11 to 23 mm wide, glabrous; lateral veins 5 to 6 pairs, with transverse veinlets.

Distributed in Yunnan, Fujian Provinces. Tender.

2. *Sinobambusa nephroaurita* C. D. Chu et C. S. Chao, sp. nov.

Culms 5 m tall, 2.5 cm in diameter. Young culms slightly pruinose, white, barbed. Sheath scar puberulent at first, becoming corky and swollen later. Mid-internodes 45 cm long, with prominent longitudinal lines. Internodes between branching nodes compressed or grooved. Primary branches 3 to many at midculm nodes. Twigs compressed, bearing 4 to 5 leaves. Leaf sheaths longitudinal, ciliate, glabrous. Leaf auricles small or none. Leaf blades lanceolate, of thin texture, 10 to 19 cm long, 14 to 23 mm wide, glabrous, serrulate, lateral veins 4 to 6 pairs, transverse veinlets on both surfaces.

Distributed in Guangdong, Guangxi, and Sichuan Provinces. Tender.

3. *Sinobambusa incana* Wen sp. nov.

Culms 4 to 6 m tall, 2 to 3 cm in diameter. Culms yellowish-green, glabrous, zigzag, lineate. Internodes nearly half cylindrical. Nodes prominently swollen, shining. Sheath scars corky, puberulent, but glabrate later. Leaves 2 to 4 on twigs. Leaf sheaths glabrous, lineate, ciliate. Leaf auricles none. Leaf blades lanceolate, 5 to 8 cm long, 7 to 10 mm wide, serrulate, glabrous, transverse veinlets prominent.

Distributed in Guangdong Province. Tender.

This sp. is similar to *S. tootsik;* but sheath auricles very small, pillow-like, not protruding; culm sheath puberulent only at base; leaf blades smaller and glabrous beneath.

4. *Sinobambusa tootsik* (Sieb.) Makino

Rhizomes amphipodial; culms scattered, 7 m tall, 3.5 cm in diameter. Young culms glabrous, dark green, with pruinose zone under nodes. Old culms green, lineate. Sheath scars corky, purple-setose, becoming glabrate later. Internodes between unbranched nodes cylindrical; but those between branching nodes compressed, 80 cm long. Nodal ridges more swollen than sheath scars. Primary branches 3 (rarely 5 to 7) at midculm nodes. Internodes of branches half cylindrical. Leaves 5 to 6 (rarely 3 to 9) on twigs. Leaf sheaths glabrous, lineate, with oral setae, auricles obsolescent. Leaf blades lanceolate to narrow-lanceolate, 6 to 22 cm long, 1 to 3.5 cm wide, of papery texture, green and glabrous above, grayish and puberulent beneath.

Distributed in Fujian, Guangdong, Sichuan, and Zhejiang Provinces. Hardy to −7°C.

4a. *S. tootsik* var. ***tenuifolia*** (Koidz) S. Suzuki

Leaf blades glabrous on both surfaces.

Distributed in Wuzhou of Guangxi Province.

4b. *S. tootsik* var. ***dentata*** Wen var. nov.

The top of sheath ligules serrate or double serrate, not entire. Leaf blades glabrous beneath.

Distributed in Fujian Province.

4c. *S. tootsik* var. ***laeta*** (McClure) Wen.

Sheath blades purplish. Leaf auricles and oral setae prominent.

Distributed in Guangzhou of Guangdong Province.

5. *Sinobambusa intermedia* McClure

Culms 5 m tall, 2 cm in diameter. Midculm internodes 50 to 60 cm long, not compressed (but top-culm internodes compressed), white-puberulent at first, becoming glabrate, rough feeling, lineate, light green, with farinose zone under nodes. Nodal ridges swollen. Sheath scars corky, densely bristly at first, later becoming glabrate. Primary branches 3 (sometimes 5 to 7) on midculm nodes, but only 1 to 2 on lower nodes. Twigs single on branch nodes. Leaves 2 to 4 on twigs. Leaf sheaths glabrous, with longitudinal veins, ciliate. Leaf auricles none or obsolescent. Leaf blades of firm-craft-papery texture, ovate-lanceolate, 13 to 28 mm wide, glabrous above, glabrous or hairy on the base beneath, serrulate, with transverse veinlets on both surfaces.

Distributed in Guangdong, Gungxi, Zhejiang, Yunnan, and Sichuan Provinces. Hardy to −7°C.

6. *Sinobambusa pulchella* Wen sp. nov.

Culms 4 to 6 m tall, 2 to 3 cm in diameter. Primary branches 3 on nodes, secondary branches single. Culm internodes puberulent, glabrate later, lineate, compressed from mid-internode downwards. Branch nodal ridges swollen. Branch sheaths of leathery texture on the base of twigs, late deciduous, amplexicaul. Leaves 2 to 4 on twigs. Leaf sheaths nearly glabrous, but ciliate, with longitudinal veins. Oral setae thick and erect; leaf auricles absent. Leaf blades broad-lanceolate, 10 to 17 cm long, 2 to 3.5 cm wide, glabrous above, serrulate, lateral veins 6 to 7 pairs, with transverse veinlets.

Distributed on Mount Dinghushan in Guangdong Province. Tender.

7. *Sinobambusa farinose* (McClure) Wen

Culms 5 m tall, 2 cm in diameter. Internodes nearly straight, 40 to 60 cm long on midculm. Young culms glabrous, pruinose; old culms lucid, glabrous, pruinose under nodes. Internodes between branching nodes compressed, nodal ridges prominently swollen, sheath scars corky, yellow-bristly at first, glabrate later. Primary branches 3 on midculm nodes. The central branch of each node two times longer than the others. Leaves 3 to 6 on twigs. Leaf sheaths glabrous, with prominent longitudinal veins and transverse veins. Leaf auricles obsolescent. Leaf blades of thin-papery texture, 6 to 18 cm long, 1 to 2 cm wide, lanceolate to oblong-lanceolate, nearly glabrous, lateral veins 4 to 5 pairs, with transverse veinlets beneath.

Distributed in Jiangxi, Fujian, and Zhejiang Provinces. Hardy to −7°C.

8. *Sinobambusa rubroligula* McClure

Culms 2 m tall, 8 mm in diameter. Internodes grayish-green, but turning to brownish-purple on the side exposing to the sun, glabrous or slightly puberulent. Internodes between branching nodes compressed from middle downwards, 27 cm long, nearly solid, Pruinose zone under nodes. Culm nodes swollen, sheath scars corky, with thin bristles, glabrate later. Branching low, primary branches 3 on lower nodes, then 5 to 7 on upper nodes, nearly solid. Leaves 5 to 7 on twigs. Leaf sheaths glabrous or hispidulous, ciliate. Leaf blades lanceolate to oblong-lanceolate, 9 to 22 cm long, 8 to 26 mm wide, glabrous above, puberulent beneath, serrulate, lateral veins 6 pairs, transverse veinlets prominent.

Distributed in Hainan Island of Guangdong Province. Tender.

9. *Sinobambusa glabrescens* Wen sp. nov.

Culms 2 m tall, 1 cm in diameter. Internodes green, glabrous, 30 cm long at midculm, compressed from middle downwards, pruinose zone under nodes. Nodal ridges slightly swollen, sheath scars not swollen, puberulent, glabrate later. Primary branches 3 on nodes. Leaves 3 to 4 on twigs. Leaf sheaths glabrous, with prominent longitudinal veins, ciliate. Leaf blades lanceolate, 9 to 11 cm long, 11 to 15 mm wide, serrulate, glabrous above, puberulent beneath, transverse veinlets on both surfaces, lateral veins 4 pairs.

Distributed in Fujian Province. Tender.

10. *Sinobambusa gigantea* Wen sp. nov.

Culms 17 m tall, 10 cm in diameter. Young culms pruinose, denser under nodes. Old culms yellowish-green; intranodes 1 cm wide; pruinose in and under nodes. Nodal ridges swollen, liratus. Sheath scars slightly swollen and narrow. Internodes grooved or compressed on branching side. Primary branches 3 on nodes. Leaves 3 to 4 on twigs. Leaf sheaths 45 mm long, glabrous, leaf auricles and oral setae none. Leaf blades lanceolate, 8 to 13 cm long, 14 to 20 mm wide; green and glabrous above, light green and puberulent on the base beneath. Lateral veins 5 to 6 pairs, transverse veinlets visible.

Distributed in Longquan Prefecture of Zhejiang Province and Jian-ou Prefecture of Fujian Province. Tolerates light frost.

11. *Sinobambusa puberula* Wen sp. nov.

Culms 6 m tall, 3 cm in diameter. Pale yellow bristles and prominent longitudinal lines between nodal ridges and sheath scars. Internodes compressed between branching nodes. Sheath scars corky, slightly swollen (1 mm high); nodal ridges swollen, liratus. Primary branches 3 to 5 on midculm nodes. Leaves 2 to 3 on twigs. Leaf sheath 5 to 6.5 cm long, of leathery texture, lineate. Leaf auricles prominent, but sometimes obsolete, only some setae remain. Leaf blades lanceolate, comparatively smaller but thick, 9 to 19 cm long, 10 to 18 mm wide; green and glabrous above, pale green and puberulent beneath.

Distributed in Lingui Prefecture of Guangxi Province. Tender.

12. *Sinobambusa urens* Wen sp. nov.

Culms 7 m tall, 3 cm in diameter. Internodes of midculm 50 cm long. Pruinose zone under nodes only. Culm wall thin. Nodal ridges swollen, liratus; sheath scars corky, brown-bristly at first, then glabrate. Primary branches 3 on nodes. Leaves 4 to 5 on twigs; glabrous. Leaf sheaths 65 mm long, glabrous, with prominent longitudinal lines, smooth on edge. Leaf blades lanceolate to broad-lanceolate, of thin texture, glabrous, lateral veins 5 to 8 pairs, transverse veinlets prominent.

Distributed in Fuling Prefecture of Fujian Province. Tolerates light frost.

13. *Sinobambusa scabrida* Wen sp. nov.

Culms 5 m tall, 15 to 20 mm in diameter. Internodes glabrous, with prominent longitudinal lines. Culms pruinose at earlier stage, denser between and under nodal ridges and sheath scars. Nodal ridges swollen and liratus. Sheath scars corky, brown-tomentose at earlier stage. Bamboo shoots appear in winter. Leaves 3 to 4 on twigs. Leaf sheaths 40 to 45 mm long, glabrous, lineate; leaf auricles and oral setae none. Leaf blades lanceolate to narrow-lanceolate, 8 to 11 cm long, 11 to 18 mm wide, glabrous, serrulate on one side only, lateral veins 4 to 5 pairs, not evenly distributed on both sides of midrib; transverse veinlets prominent.

Distributed in Lingdong Prefecture of Guangxi Province. Tender.

14. *Sinobambusa henryi* (McClure) Chu et Chao

Culms straight, 7 to 13 m tall, 3 to 8 cm in diameter. Internodes green, 30 to 60 cm long, cylindrical; but grooved between branching nodes; and compressed on up-culm; slightly hairy at earlier stage, glabrate afterwards. Pruinose zone under nodes at earlier stage. Intranodes 3 to 4 mm long. Nodal ridges swollen, sheath scars corky, bristly at earlier stage. Primary branches 3 on lower nodes, somewhat spreading, then 5 to 7 from midculm up. Leaves 3 to 5 on twigs. Leaf sheaths glabrous or scattered-appressed-bristly, ciliate. Leaf ligules obsolescent. Leaf blades lanceolate to oblong-lanceolate, 8 to 15 cm long, 15 to 23 mm wide, serrulate; green and glabrous above, whitish and with scattered fine hairs beneath. Lateral veins 4 pairs, transverse veinlets on both surfaces.

Distributed in Yiyang Prefecture of Guangxi Province. Tender.

15. *Sinobambusa anaurita* Wen sp. nov.

Culms 5 m tall, 2 to 2.5 cm in diameter. Internodes 30 cm long, zigzag. Young culms white-tomentose, pruinose zone only under nodes. Nodal ridges prominently swollen; sheath scars corky, bristly fugacious, glabrate later. Intranodes 5 mm long. Primary branches 3 on nodes. Leaves 2 to 4 on twigs. Leaf sheaths pale green, with longitudinal and transverse veins, glabrous except white-pilose on apex. Leaf auricles usually obsolete. Leaf blades lanceolate to narrow-lanceolate, 9 to 16 cm long, 10 to 16 mm wide; green and glabrous above, whitish-green and puberulent beneath; serrulate. Lateral veins 4 to 5 pairs, transverse veinlets visible on both surfaces.

Distributed on Mount Jinggangshan of Jiangxi Province and Shanghang Prefecture of Fujian Province. Tolerates light frost.

16. *Sinobambusa striata* Wen sp. nov.

Culms 10 m tall, 5 cm in diameter. Internodes 65 cm long, Young culms green, pruinose zone only under nodes; erect-setulose; glabrate afterwards. Nodal ridges as swollen as or more swollen than sheath scars. Sheath scars corky, bristly at earlier stage. Leaves 2 on twigs, leaf sheaths 4 to 5 cm long, with longitudinal lines, glabrous; leaf auricles and oral setae none. Leaf blades lanceolate, 9 to 14 cm long, 12 to 18 mm wide; green and glabrous above, whitish-green and puberulent beneath. Lateral veins 4 to 6 pairs, transverse veinlets visible.

Distributed in Mount Jinggangshan of Jiangxi Province. Tolerates light frost.

Path through a bamboo grove in Hanzhou.

XVII. *Indosasa* McClure

Tall bamboos, rhizomes monopodial or amphipodial. Culms scattered, erect, grooved on branching side. The length of grooves up to ½ or more of the length of internodes. Nodal ridges prominent, geniculate. Primary branches 3 on midculm nodes. The central branches thicker, the side ones thinner. Culm sheaths deciduous, of leathery or thin-leathery texture, usually bristly but not maculate. Sheath blades triangular or triangular-lanceolate, erect or reflexed. Transverse veinlets distinct. Veins tesselate-reticulate.

KEY to the Spp. of *Indosasa* Native to China

1a. Culm sheath auricles absent. 2
1b. Culm sheath auricles remain. 8
2a. Culm sheaths distinct hairy, oral setae a few. 3
2b. Culm sheaths nearly glabrous, oral setae none. Sheath blades glabrous on both surfaces, smooth. . . . . (1) *I. glabrata*
3a. Both nodal ridges and sheath scars prominently swollen. Bristly on abaxial sheath surfaces. . . . 4
3b. Nodal ridges and sheath scars slightly swollen. Culm sheath sparse-bristly. 7
4a. Culm sheaths abscising late or early. Sheath blades hairy or scabrous. 5
4b. Culm sheaths abscising early. Sheath blades glabrous. 6
5a. Sheath blades hispidulous, scabrous. Leaves usually 3 to 4 on twigs. . . . . . (2) *I. crassiflora*
5b. Sheath blades glabrous on abaxial surfaces, scabrid on adaxial surfaces. Leaves (4) 5 to 9 (10) on twigs. . . . . (3) *I. ingens*
6a. Sheath blades erect. Internodes without longitudinal ribbed lines. Leaves 4 to 7 on twigs. . . . . (4) *I. purpurea*
6b. Sheath blades recurved. Internodes with longitudinal ribbed lines. Leaves 3 to 5 on twigs. . . . . (5) *I. triangulata*
7a. Leaf blades large, 11 to 28 cm long, 1.5 to 5 cm wide; sparse-hispidulous beneath. . . . . (6) *I. angustata*
7b. Leaf blades small, 5 to 17 cm long, 1.2 to 2.5 cm wide, glabrous. . . . (7) *I. spongiosa*
8a. Usually 1 leaf on twigs (seldom 2), leaf sheaths clasping. Culm sheaths sparse-bristly (later shedding). . . . . (8) *I. shibataeoides*
8b. Leaves 2 to many on twigs. 9
9a. Nodal ridges and sheath scars prominently swollen, geniculate. Culm sheath tufted-bristly, not ciliate. 10
9b. Nodal ridges and sheath scars moderately swollen, not geniculate. Culm sheath scattered-bristly, ciliate. 12
10a. Young culms dense-pruinose. Culm sheath auricles small. Leaf blades large, usually 11 to 22 cm long, 1.5 to 3 cm wide. . . . . (9) *I. sinica*
10b. Young culms not pruinose. Culm sheath auricles semiorbicular. Leaf blades smaller. 11
11a. Leaves 4 to 7 on twigs. Leaf blades linear-lanceolate to lanceolate, 6 to 14 cm long, 1 to 1.5 cm wide, lateral veins 3 to 4 pairs. . . . . (10) *I. parvifolia*
11b. Leaves 2 to 4 on twigs. Leaf blades lanceolate, 8 to 15 cm long, 1 to 2.5 cm wide, lateral veins 4 to 5 pairs. . . . (11) *I. lipoensis*
12a. Young culms sparse short hairy under nodes only. Culm pith thickened with irregular strata. Culm sheath auricles prominent, nearly falciform. Sheath blades broad-lorate-lanceolate, smooth on both surfaces. . . . . (12) *I. lingchuanensis*

12b. Young culms dense-barbed, scabrous. Culm pith transversely dissepiment-like and sponge-like incrassate. Culm sheath auricles smaller. Sheath blades triangular to triangular-lanceolate, scabrous on both surfaces. 13

13a. Young culms dense-pruinose. Branches fastigiate. Leaf blades 9 to 12 cm long, 1.2 to 2.6 cm wide, glabrous. Culm sheath ligules acute at apex. . . . . (13) *I. longispicata*

13b. Young culms not pruinose. Branches spreading nearly horizontal. Leaf blades 13 to 25 cm long, 2 to 4 cm wide, sparse-white-hairy on underside. Culm sheath ligules truncate or slightly arcuate at apex. . . . . (14) *I. patens*

1. ***Indosasa glabrata*** C. D. Chu et C. S. Chao, sp. nov.

Culms 3 m tall, 2 cm in diameter. Young culms green glabrous; pruinose only on nodes. Pith of culms ring-form. Old culms pale green. Internodes 20 to 30 cm long. Nodes prominently geniculate. Primary branches 3 on nodes. Branch nodes geniculate. Leaves 2 to 4 on twigs. Leaf auricles and oral setae glabrate. Leaf blades elliptic-lanceolate to lanceolate, 11 to 23 cm long, 2 to 5 cm wide, glabrous, lateral veins 5 to 7 pairs.

Distributed in Guangxi Province. Tender.

2. ***Indosasa crassiflora*** McClure

Culms 4 to 5 m tall, 2 to 4 cm in diameter. Culm wall thick, nearly solid on lower culm. Pith cavity small. Nodal ridges prominently swollen. Young culms glabrous, branches and leaves sparse. Primary branches 2 to 3 on nodes, sometimes single. Leaves usually 3 to 4 on twigs. Leaf sheaths shedding easily. Oral setae tufted. Leaf ligules short, setulose on back. Leaf blades large, lanceolate to oblong-lanceolate, 10 to 25 cm long, 1.7 to 4.5 cm wide, pruinose beneath. Lateral veins 5 to 8 pairs.

Distributed in Dongzing Prefecture of Guangdong Province. Tender.

3. ***Indosasa ingens*** Hsueh et Yi, sp. nov.

Culms 2 to 6 m tall, (1) 3 to 5 cm in diameter, erect; rhizomes amphipodial. Approximately 15 internodes on culms. Internodes (15) 30 to 40 (65) cm long, green or purplish green, yellowish brown-appressed-setose on upper culm; scabrous, pruinose zone under nodes. Culm wall thick. Nodal ridges slightly swollen; but prominently swollen on branching nodes, corky, yellowish-brown, somewhat smooth, glabrous. Primary branches (1) 3 (5) on nodes, fastigiate; internodes triangular or square at base, sparse-bristly or glabrous; branch nodes prominent. Leaf sheaths 5 to 12 cm long, yellowish-brown-fimbriate-ciliate. Leaf auricles none, but only 2 to 3 setae remain or none. Leaf ligules truncate at apex, glabrous, 1 mm long. Leaf blades lanceolate to oblong-lanceolate, of thick-papery texture, glabrous, (7) 12 to 24 cm long, (1.2) 2.5 to 4.5 cm wide; acuminate or tail-like acuminate at apex; cuneate or broad-cuneate at base, serrulate, slightly scabrous. Lateral veins 6 to 7 pairs, transverse veinlets distinct.

Distributed at an altitude of 950 to 1600 m in Yunnan Province. Tender.

4. ***Indosasa purpurea*** Hsueh et Yi, sp. nov.

Culms 3 to 10 m tall, 2 to 8 cm in diameter, erect to the top. Internodes (10) 30 to 45 (75) cm long, glabrous, pruinose zone under nodes. Culm pith sponge-like. Approximately 20 to 22 internodes on culms. Young culms brown-erect-bristly only on sheath scars. Nodal ridges swollen to very swollen and liratus. Leaves 4 to 7 on twigs. Leaf sheaths glabrous. Leaf auricles obsolescent. Oral setae absent or seldom 2 to 3, yellowish-brown. Leaf blades lanceolate, of papery texture, 12 to 21 cm long, 16 to 26 mm wide; glabrous or sparse-short-hairy beneath; serrulate on one side only. Lateral veins (3) 5 to 7 pairs, transverse veinlets distinct. Leaf stalks (2) 3 to 8 mm long, glabrous.

Distributed at an altitude of 1100 to 1650 m in Yunnan Province. Tender.

5. ***Indosasa triangulata*** Hsueh et Yi, sp. nov.

Culms 3 to 5 m tall, 1 to 2.5 cm in diameter, rhizomes amphipodial. Internodes (10) 30 (40) cm long, cylindrical, but slightly swollen on the branching side, green, glabrous; slightly pruinose when young; longitudinal-ribbed-lineate. Sheath scars prominently swollen, corky. Culm sheaths yellowish-brown-hairy at base. Nodal ridges swollen, or prominently swollen on branching nodes, glabrous, sometimes black farinose. Leaves

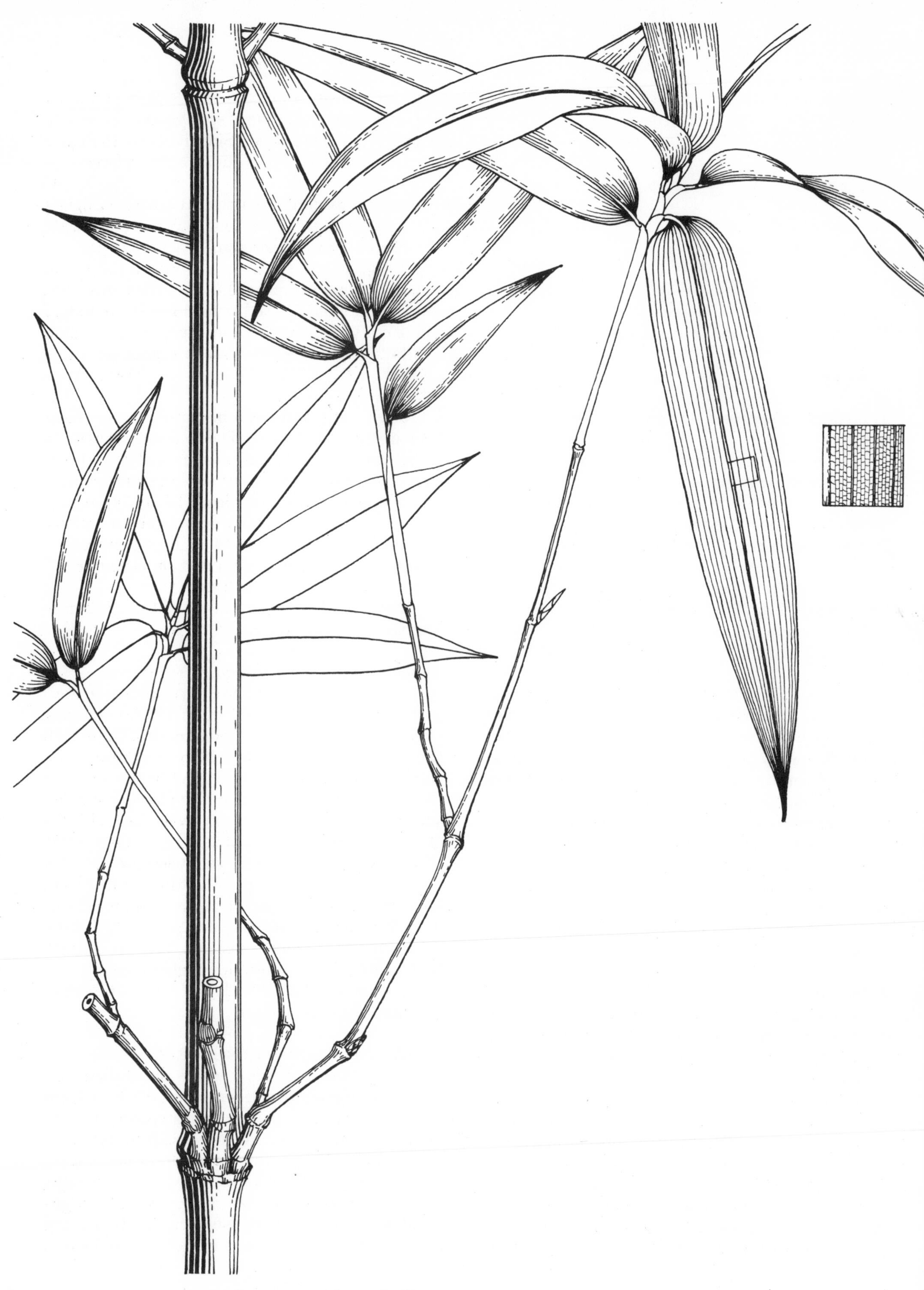

Indosasa crassiflora McClure

3 to 5 on twigs. Leaf sheaths 5.5 to 7 cm long. Leaf auricles none. Leaf ligules arcuate or truncate at apex, purple. Leaf blades lanceolate to broad-lanceolate, of firm-papery texture, glabrous; 9 to 19 cm long, 1.2 to 2.5 cm wide; green above, grayish-white beneath, acuminate at apex; cuneate to broad-cuneate at base; serrulate, scabrous. Lateral veins 5 to 7 pairs, transverse veinlets slightly distinct. Leaf stalks 3 to 8 mm long.

Distributed at an altitude of 1200 m in Yunnan Province. Tender.

6. *Indosasa angustata* McClure

Culms 14 m tall, 10 cm in diameter. Nodal ridge slightly swollen. Culm sheaths of thin-leathery texture, narrow and long, acuminate to the apex; pale green, turning to pale brown when dry; veins distinct, bristly between veins. Sheath auricles obsolescent, several erect oral setae, 1 to 1.5 cm long. Leaf blades large, grayish-green beneath, sparse-hispidulous, scabrous.

Distributed in a small area on Mount Daqingshan of Guangxi Province. Tender.

7. *Indosasa spongiosa* C. S. Chao et B. M. Yang, sp. nov.

Culms 5 to 8 m tall, 1 to 6 cm in diameter, rhizomes monopodial. Culm pith sponge-like. Internodes 20 to 35 cm long on midculm, grooved at branching side, slightly scabrous. Nodes slightly swollen, geniculate. Pruinose under nodes when young. Primary branches 3 on nodes, sometimes only 1 to 2 on a few of the lowest nodes; spreading. Leaves 3 to 5 on twigs. Leaf sheaths glabrous. Leaf auricles obsolescent or very small. Oral setae none or only a few, white, crooked. Leaf blades lanceolate to oblong-lanceolate, 5 to 17 cm long, 12 to 25 mm wide; acuminate at apex; cuneate at base; serrulate on both sides, or one side entire; glabrous. Lateral veins 5 to 6 pairs, reticulate veins in square pattern. Leaf stalk 3 to 5 mm long.

Distributed in Hunan Province. Ornamental. Tolerates light frost.

8. *Indosasa shibataeoides* McClure

Culms 15 m tall, 10 cm in diameter. Culm sheaths brownish-purple or pale orange-red, but becoming green on upper culm; blackish-brown-striate; sparsely-pruinose and bristly (shedding easily). Culm sheaths on small culms glabrous. Sheath auricles moderately large in size, oral setae short, 2 to 3 mm long. Sheath blades green, with distinct purple veins. Leaf blade large, elliptic-lanceolate, whitish-green beneath.

Distributed in Guangzi, Guangdong, and Hunan Provinces. Tender.

9. *Indosasa sinica* C. D. Chu et C. S. Chao, sp. nov.

Culms 10 m tall, 6 cm in diameter. Young culms green, dense-pruinose, sparse-hispidus, slightly scabrous. Old culms pale brown or dark green. Internodes 35 to 50 cm long. Nodes prominently geniculate. Culm wall thick. Primary branches 3 at nodes, spreading. Branch nodes geniculate. Leaves 3 to 9 on twigs. Oral setae pale purple, 8 mm long, shedding easily. Leaf blades linear-lanceolate, green on both sides. Lateral veins 5 to 6 pairs.

Distributed in Guangxi and Yunnan Provinces. Tender.

10. *Indosasa parvifolia* C. S. Chao et Q. H. Dai, sp. nov.

Culms 6 m tall, 3.5 cm in diameter. Young culms dark green, dense-white-bristly; scabrous; pruinose on nodes only. Old culms green. Internodes 25 to 40 cm long. Nodes prominently geniculate. Culm wall thick. Primary branches 3 on nodes (rarely 1 to 2). Leaf blades smaller, glabrous.

Distributed in Guangxi Province. Tender.

11. *Indosasa lipoensis* C. D. Chu et K. M. Lan, sp. nov.

Culms 10 m tall, 3 to 4 cm in diameter. Young culms dense-setulose, somewhat scabrous. Culm pith sponge-like. Internodes 30 to 40 cm long, distinctly grooved on the lower part. Nodal ridges swollen. Sheath scars glabrous. Primary branches 3 on nodes, spreading. Branch nodes prominently swollen, geniculate. Leaf sheaths glabrous. Leaf auricles small, sparse-erect-hairy, shedding easily. Leaf blades glabrous, serrulate.

Distributed in Guizhou Province (Libo Prefecture). Tender.

12. *Indosasa lingchuanensis* C. D. Chu et C. S. Chao, sp. nov.

Culms 4 m tall, 3 cm in diameter, green, sparse-bristly, slightly scabrous, hollow. Culm wall irregularly thickened inside. Internodes 30 to 40 cm long. Nodes and sheath scars distinct. Primary branches 3 on nodes. Leaves 3 to 5 on twigs. Leaf sheaths ciliate. Leaf auricles small, oral setae erect. Leaf blades smaller, linear-lanceolate, 6.5 to 14 cm long, 1 to 2.3 cm wide, glabrous. Lateral veins about 5 pairs.

Distributed in Guangxi Province. Tender.

13. *Indosasa longispicata* W. Y. Hsiung et C. S. Chao, sp. nov.

Culms 10 to 15 m tall, 6 cm in diameter. Young culms green, dense-white-barbed, scabrous. Culm pith sponge-like. Old culms yellowish-green. Internodes 40 to 50 cm long. Nodes slightly swollen, not geniculate. Primary branches 3 (rarely 5) on nodes. Leaves 3 to 5 on twigs. Leaf sheaths ciliate. Leaf auricles and oral setae remain. Leaf ligules obsolescent. Leaf blades smaller, linear-lanceolate to lanceolate.

Distributed in Guangxi Province. Tolerates partial shade. Tender.

14. *Indosasa patens* C. D. Chu et C. S. Chao, sp. nov.

Culms 12 m tall, 8 to 10 cm in diameter. Young culms green, purple-striate, dense-white-bristly, scabrous. Internodes 40 to 60 cm long. Nodes moderately swollen, not geniculate. Primary branches 3 (rarely 1) on nodes, spreading nearly horizontally. Leaves 2 to 5 on twigs. Leaf sheaths glabrous. Leaf auricles small, oral setae sparse, 5 to 10 mm long. Leaf blades lorate-lanceolate.

Distributed in Guangxi Province. Tolerates partial shade. Tender.

XVIII. *Fargesia* Franchet

Shrub-like bamboos. Rhizomes sympodial with elongated rhizome necks. Culms pluricespitose and also diffuse. Internodes cylindrical. Primary branches 3 to many on nodes, clustered. Culm sheaths persistent or late deciduous. Sheath scars swollen. Sheath blades slender. Sheath auricles often obscure. Leaf blades comparatively small.

Distributed at an altitude of 1000 to 3300 m in southern and southwestern China. Hardy and humidity-loving.

KEY to the Spp. of *Fargesia* Native to China

1a. Culms solid, or a few nodes at culm base nearly solid. 2
1b. Culms fistulose. 3
2a. Culms and branches solid. Upper part of internodes sparsely grayish-white setulose, prominently longitudinal-elevated-lineate. Sheath scars swollen, dark brown, glabrous; sheath callus exserted. Leaf auricles none. .... (1) *F. farcta*
2b. A few nodes and internodes at culm base nearly solid. Internodes glabrous, indistinctly longitudinal-elevated-lineate. Sheath scars swollen to prominently swollen, yellowish-brown-spiny when young. No sheath callus. Leaf auricles very small. .... (2) *F. ampullaris*
3a. Culm sheaths setose on abaxial surface. 4
3b. Culm sheaths glabrous on abaxial surface. Sheath auricles and oral setae absent, or a few setae remain. 9
4a. Sheath auricles distinct in general, but a part of the culms sometimes without sheath auricles. 5
4b. Sheath auricles obscure. 7
5a. Culm sheaths maculate. .... (3) *F. ferax*
5b. Culm sheaths not maculate. 6
6a. Culms 4 to 6 m tall, glabrous; densely pruinose when young, often becoming blackish-dirty later. Sheath proper yellowish-brown, sparsely yellowish-brown-setose on abaxial surface, not ciliate. .... (4) *F. glabrifolia*
6b. Culms 8 m tall, glabrous, not pruinose. Sheath proper densely yellowish-brown-setose on abaxial surface, also yellowish-brown-setose on rims. .... (5) *F. grossa*
7a. Culm sheaths persistent or late deciduous. 8
7b. Culm sheaths abscising early. Culms 4 m tall. Young culms green, pruinose, glabrous. Old culms purplish-brown. Sheath proper pale-brown-setose. .... (6) *F. spathacea*
8a. Culms 3 to 5 (7) m tall, green, glabrous, not pruinose. Culm sheaths persistent, as long as internodes, orange-red or gray, densely brown-setose on

abaxial surface and upper rim. .... (7) *F. hainanensis*

8b. Culms 1 to 7 m tall, green, slightly pruinose. Upper part of internodes brown- or grayish-brown-setose and densely girdled under nodes when young. Culm sheaths persistent or late deciduous, slightly longer than internodes; densely brownish-yellow- to brown-acropetal setose; brown-ciliate or not. .... (8) *F. setosa*

9a. Culm sheaths indumentum on adaxial surface; yellowish-brown or somewhat purplish; grayish-white- or grayish-yellow-setulose on the narrow part of adaxial surface; yellowish-brown to brown setulose on upper rim. .... (9) *F. semiorbiculata*

9b. Culm sheaths not hairy on adaxial surface. 10

10a. Sheath callus remains. 11

10b. No sheath callus. 12

11a. Culms 2 to 6 m tall. Nodal ridges swollen. Culm sheaths rectangular; contracted at apex as short-neck-like; both corners at base decurrent to auricle-like appendages; pale-yellow-ciliate on upper rim. .... (10) *F. collaris*

11b. Culms 3 to 4 (5.5) m tall. Nodal ridges swollen on one side only, geniculate. Culm sheaths rectangular or triangular-rectangular, pruinose on abaxial surface, yellowish-brown-ciliate. .... (11) *F. crassinoda*

12a. Culms 4 to 6.5 m tall, green and slightly pruinose when young, turning to yellow when old. Sheath scars swollen, gray, glabrous. Nodal ridges slightly swollen, glabrous. Bamboo shoot sheaths not appressed on upper part, but spreading trumpet-like. .... (12) *F. extensa*

12b. Culms 3 to 3.5 m tall, densely pruinose when young, becoming blackish-dirty later. Sheath scars swollen, gray to brown, densely setulose when young. Nodal ridges swollen, purple and glabrous when young. Bamboo shoot sheaths appressed on upper part, not spreading trumpet-like. .... (13) *F. gyirongensis*

1. *Fargesia farcta* Yi, sp. nov.

Culms 2 to 3.5 m tall, 5 to 15 mm in diameter, about 20 nodes on culm. Internodes 22 to 28 cm long, the longest 32 to 35 cm; cylindrical or slightly compressed at base on the branching side; green; pruinose when young. Nodal ridges slightly swollen to swollen, green glabrous. Intranodes green; slightly pruinose when young. Culm sheath late deciduous, of leathery texture, yellowish-brown, longer than internodes. Sheath auricles very small, oral setae several, curved, yellowish-brown, 3 to 5 mm long. Sheath blades reflexed, fugacious. Primary branches (1) 3 to 10 clustered on nodes, appressed to the culm at base, upright or fastigiate. Leaves 2 to 5 on twigs. Leaf sheaths pale green, slightly pruinose, glabrous, longitudinal ridge and veins distinct on upper part, densely grayish-brown ciliate. Leaf blades narrow-lanceolate, papery, 4 to 7.5 cm long, 5 to 8 mm wide; acuminate at apex, broad-cuneate or rarely rounded at base; green and white pilosulose above, pale green and white-pubescent (especially dense at base) beneath; serrulate and scabrous on rim. Lateral veins 2 (rarely 3) pairs. Transverse veinlets distinct.

Distributed in Xizang (Tibet) Autonomous Region. Hardy to −5°C.

2. *Fargesia ampullaris* Yi, sp. nov.

Culms 2 to 5.4 m tall, 7 to 15 mm in diameter. 26 to 34 nodes on a culm. Internodes 20 to 25 cm long, the longest 30 cm, cylindrical; densely pruinose when young. Nodal ridges swollen, glabrous, densely pruinose when young. Intranodes glabrous, less pruinose. Culm sheaths abscising late, pale yellowish-brown, leathery, asymmetrically contracted at upper part to bottleneck-like; sparsely brown-setose on upper part and also densely yellowish-brown- to brownish-black-setose at base of abaxial surface; pale-yellow to brown-setulose on the narrowed part (denser near top) of adaxial surface. Sheath auricles obscure, no oral setae.

Sheath ligules prominent, 1.5 to 5 mm long, densely yellowish-brown setulose. Sheath blades linear-lanceolate, reflexed. Primary branches many, slender, of even thickness, clustered on nodes. Leaves 4 on twigs. Leaf blades narrow-lanceolate, papery, 7 to 11 cm long, 6 to 10 mm wide, long-acuminate at apex, cuneate at base; glabrous above, sparsely gray-puberulent at base beneath, serrulate. Lateral veins 3 (rarely 2) pairs. Transverse veinlets indistinct.

Distributed in Xizang (Tibet) Autonomous Region. Hardy to −5°C.

3. ***Fargesia ferax*** (Keng) Yi

Culms 2.6 m tall, 1 cm in diameter. Primary branches many, in half-whorl pattern, on nodes, to or over 30 cm long. Branch internodes slender, 1.5 mm in diameter, 11 cm long. Leaf sheaths, stramineous. Oral setae a few, 1.5 to 3.5 mm long, yellow, fimbriate. Leaf ligules truncate at apex. Leaf blades 2 to 6 cm long, 3 to 5 mm wide, tensile, acuminate at apex, narrowed to short stalk at base; pale green and pilosulose at base beneath, slightly serrulate on one side only, nearly smooth on the other side. Lateral veins 2 to 3 pairs. Transverse veinlets distinct.

Distributed at an altitude of 2400 m in Sichuan Province. Hardy to −15°C.

4. ***Fargesia glabrifolia*** Yi, sp. nov.

Culms 4 to 6 m tall, 8 to 20 mm in diameter. Nodes 20 to 34 on a culm. Internodes 30 to 35 cm long, the longest reaching 46 cm; those near culm base 8 to 9 cm long; cylindrical, glabrous, longitudinal-elevated-lineate. Sheath scars not swollen or slightly swollen, glabrous, often dirty. Nodal ridges not swollen, glabrous, lucid. Primary branches many clustered on every node, spreading at an angle of 45°, of even thickness. Bamboo shoot sheath purplish-red, becoming yellowish-brown when dry; sparsely grayish-white to grayish-yellow-setose, shedding later; pale-yellowish-brown-setose on rim. Culm sheaths late deciduous, coriaceous, triangular-rectangular. Sheath auricles not evident on lower culm sheaths, but small sheath auricles found on upper-culm sheaths. Sheath ligules truncate, purple. Sheath blades linear-lanceolate, papery, (3.5) 5 to 8 (9) cm long, 4 to 5 mm wide; acuminate at apex, broad-cuneate at base, glabrous, indistinctly serrulate. Transverse veinlets indistinct.

Distributed in Xizang (Tibet) Autonomous Region. Hardy to −5°C.

5. ***Fargesia grossa*** Yi, sp. nov.

Culms 8 m tall, 2 to 3.5 cm in diameter. 25 to 38 nodes on culm. Internodes cylindrical, 15 to 45 cm

Tall bamboos grow on a beautiful island in the very heart of Guilin City. In the background are peaks typical of the Guilin area.

long, longitudinal-elevated lineate, not pruinose, glabrous, somewhat lucid. Sheath scars swollen, glabrous, scabrid, often with sheath callus. Nodal ridges not swollen or slightly swollen, pale green, glabrous. Primary branches many clustered on nodes, spreading at an angle of 25° to 30°. Culm sheaths coriaceous, long-triangular. Culm sheaths on lower culms without sheath auricles. Sheath ligules truncate. Sheath blades linear-lanceolate, (2) 4.5 to 8 (11) cm long, 5 to 8 mm wide; acuminate at apex, cuneate at base; green and glabrous above, pale green and pubescent at midrib base beneath; serrulate. Lateral veins (2) 3 (4) pairs, transverse veinlets indistinct.

Distributed in Xizang (Tibet) Autonomous Region. Hardy to −5°C.

6. *Fargesia spathacea* Franchet
[*Sinarundinaria nitida* (Mitf.) Nakai]

Culms 3 m tall, 1 to 1.5 cm in diameter. Internodes 6 to 8 cm long. Sheath scars prominently swollen, with sheath callus. Nodal ridges not swollen. Culm sheaths abscising early, brownish, stramineous when dry. Sheath auricles and oral setae absent. Sheath blades reflexed. Sheath ligules short, arcuate at apex, ciliolate. Primary branches 3 to many on nodes, slender. Leaves 2 to 3 on twigs. Leaf sheaths purple, yellow-ciliate at apex. Leaf auricles indistinct. Leaf blades 6 to 13 cm long, 5 to 14 mm wide. Lateral veins 4 pairs.

Distributed at an altitude of 1000 to 3000 m in northern Sichuan, Yunnan, Hubei, Jiangxi, Gansu, Shaanxi, and Hunan Provinces. Hardy to −20°C.

7. *Fargesia hainanensis* Yi, sp. nov.

Culms 3 to 5 (7) m tall, (1) 2 to 3.5 (5) cm in diameter. 25 to 29 (35) nodes on a culm. Internodes 24 to 28 (32) cm long, the shortest at culm base 2 to 8 cm long only; cylindrical, sometimes slightly compressed or with a few longitudinal ridges and grooves on the branching side; prominently longitudinal-elevated lineate. Sheath scars not swollen or slightly swollen, glabrous. Nodal ridges slightly swollen or not, also glabrous. Intranodes glabrous, slightly lucid. Bamboo shoots purplish-red, densely brown setose. Culm sheaths coriaceous. Sheath ligules arcuate at apex. Sheath blades linear-lanceolate, reflexed, glabrous, serrulate. Primary branches 3 to 7 clustered on every node of culms, spreading at an angle of 30° to 40°, glabrous, longitudinal-elevated-lineate. Leaves (2) 3 to 5 (6) on twigs. Oral setae on leaf sheaths several, 3 to 5 mm long, gray, erect or slightly curved. Leaf blades narrow-lanceolate, papery, 4 to 12 cm long, 5 to 9 mm wide, long-acuminate at apex, cuneate at base, green and glabrous above; grayish-green, glabrous or pilosulose near base beneath; serrulate. Lateral veins 3 to 4 pairs, transverse veinlets distinct.

Distributed at an altitude of 1560 to 1800 m in Hainan Island of Guangdong Province. Tender.

8. *Fargesia setosa* Yi, sp. nov.

Culms 1 to 7 m tall, 5 to 35 mm in diameter, curving slightly at top. Nodes 16 to 32 in a culm. Internodes on midculm 18 to 28 cm long, the longest 53 cm, the shortest at culm base 8 to 18 cm long; cylindrical. Sheath scars prominently swollen, brown-setose when young, glabrate later. Nodal ridges slightly swollen, glabrous. Primary branches 3 to 7 clustered on nodes, of about even thickness, somewhat purplish when young, glabrous, pruinose, at an angle of 25° to 30° to the culm. Sheath proper from middle downwards coriaceous, while the rims and the upper part papery. Sheath auricles obscure. Sheath ligules truncate at apex. Sheath blades triangular-linear to linear lanceolate, reflexed, green, fugacious. Leaves 3 to 5 on twigs. Leaf blades lanceolate, papery, 4 to 17 cm long, 4 to 18 mm wide, acuminate at apex, broad cuneate at base; scabrid above, grayish-brown pilosulose beneath, densely gray- or grayish-brown pubescent near base on both surfaces; serrulate and scabrous on rims. Lateral veins 2 to 3 pairs. Transverse veinlets merely visible.

Distributed in Xizang (Tibet) Autonomous Region. Hardy to −10°C.

9. *Fargesia semiorbiculata* Yi, sp. nov.

Culms 4.2 m tall, 6 to 13 mm in diameter. Nodes 24 on a culm. Internodes 15 to 20 cm long, the longest 29 cm, while the shortest near culm 2.5 to 8 cm long; cylindrical, glabrous; densely pruinose when young, becoming blackish-dirty afterwards; longitudinal-elevated-lineate not distinct. Sheath scars slightly swollen, glabrous. Nodal ridges swollen, glabrous, lucid. Primary branches many clustered on nodes, slightly geniculate-curved, slender, slant-spreading, of even thickness; pruinose when young, becoming blackish-dirty later. Sheath proper coriaceous except upper part of thinner texture, rectangular or rarely long-triangular. Sheath blades triangular to linear-lanceolate, often reflexed. Leaves 2 to 4 (5) on slender twigs. Leaf blades narrow-lanceolate, papery, (2) 5 to 8.5 (10) cm long, 4 to 5 mm wide; acuminate at apex, cuneate at base, serrulate and scabrid on rims, glabrous. Lateral veins 2 to 3 pairs, transverse veinlets indistinct.

Distributed in Xizang (Tibet) Autonomous Region. Hardy to −5°C.

10. ***Fargesia collaris*** Yi, sp. nov.

Culms 2 to 6 m tall, 1 to 3.5 cm in diameter. Internodes 17 to 28 cm long; densely pruinose when young, glabrous, lucid, indistinctly longitudinal-elevated-lineate. Sheath scars swollen, with sheath callus, glabrous. Nodal ridges swollen, glabrous. Internodes densely pruinose when young. Primary branches many clustered on every node, glabrous, slightly lucid, with pruinose zone under nodes. Culm sheaths coriaceous. Leaves 3 to 4 on twigs. Leaf ligules ligulate-protruding. Leaf blades lanceolate, 2 to 12.5 cm long, 3 to 19 mm wide, of thin texture, dark green above, pale green and scabrid beneath, acuminate at apex, rounded or broad-cuneate at base, serrulate and scabrous on rims.

Distributed in Xizang (Tibet) Autonomous Region. Hardy to −5°C.

11. ***Fargesia crassinoda*** Yi, sp. nov.

Culms 3 to 4 (5.5) m tall, 10 to 20 mm in diameter. Nodes 23 to 30 on a culm. Internodes 15 to 18 cm long, the longest 21 cm, while the shortest at culm base 4 to 7 cm long only, cylindrical or oblate, slightly compressed at base on branching side, sometimes with longitudinal ridges and longitudinal grooves. Young culms densely pruinose, glabrous, becoming purplish-green or yellowish-green when old, lucid and smooth. Sheath scars swollen, grayish-brown, glabrous, sometimes with sheath callus. Nodal ridges usually on one side only, geniculate, glabrous. Intranodes densely pruinose when young, glabrous. Culm sheaths abscising early, coriaceous, pale gray, rectangular or triangular-rectangular. Sheath blades triangular or linear-lanceolate, erect, gray or grayish-brown, aristate-serrulate. Primary branches (1) 3 to 6 clustered on nodes, fastigiate or slant-spreading, of nearly even thickness, slightly zigzag, glabrous, pruinose. Leaves (2) 3 to 5 (6) on twigs. Leaf blades narrow-lanceolate, papery, (4.5) 6 to 7 (9) cm long, 5 to 8 (10) mm wide; acuminate at apex, broad-cuneate or nearly rounded at base; green above, grayish-green beneath; glabrous, serrulate and scabrous at rims. Lateral veins 2 pairs. Transverse veinlets distinct.

Distributed in Xizang (Tibet) Autonomous Region. Hardy to −5°C.

12. ***Fargesia extensa*** Yi, sp. nov.

Culms widely separated, 4 to 6.5 m tall, 1 to 2.8 cm in diameter, upright or slightly curving at the top. Nodes 25 to 32 on a culm. Internodes cylindrical, slightly compressed on branching side, 20 cm long, the longest 32 cm, while the shortest at culm base are only 10 cm long. Branching usually from the 9th to 13th nodes from culm base. Primary branches (3) 5 (7) on nodes, spreading, purplish when young, glabrous, lucid. Culm sheaths coriaceous from middle to the base, papery on upper part. Sheath ligules tridentate at apex, membranous, purplish-brown, glabrous. Leaves 3 to 4 (8) on twigs. Leaf blades papery, narrow-lanceolate, 5.5 to 16.5 cm long, 7 to 14 mm wide, acuminate at apex, cuneate at base, green and lucid above, pale-green beneath, glabrous, serrulate and scabrous at rims.

Distributed in Xizang (Tibet) Autonomous Region. Hardy to −10°C.

13. ***Fargesia gyirongensis*** Yi, sp. nov.

Culms 3 to 3.5 m tall, 1 to 1.2 cm in diameter. Nodes 22 to 26 on a culm. Internodes on midculm 15 to 19 cm long, on culm base 3.5 to 5.5 cm long; cylindrical, the base on branching side also cylindrical; distinctly or indistinctly longitudinal-elevated-lineate. Culm sheaths abscising early, coriaceous. Primary branches many clustered on nodes, slender, of nearly even thickness, slant-spreading, pruinose zone under nodes. Leaves 4 to 6 on twigs. Leaf blades lanceolate, papery, (3.5) 7 to 11 (14) cm long, 6 to 12 mm wide; acuminate at apex, broad-cuneate or rarely rounded at base; green and glabrous above, pale green and grayish-white pubescent along the base of midrib beneath; serrulate and scabrous at rims. Lateral veins (2) 3 (4) pairs. Transverse veinlets obscure.

Distributed in Xizang (Tibet) Autonomous Region. Hardy to −5°C.

XIX. *Yushania* Keng f.

Sub shrub-like or shrub-like bamboos. Rhizomes sympodial, with elongated rhizome necks. Culms erect, pluricespitose and also diffuse. Internodes cylindrical or slightly compressed at base on branching side. Culm sheaths late deciduous or persistent, of thin texture. Primary branches 3 to 7 on every node of culms, rarely less or many, upright or fastigiate.

KEY to the Spp. of *Yushania* Native to China

1a. Culm internodes glabrous. 2
1b. Culm internodes sparsely setulose when young. 3
  2a. Culm sheaths pale yellow, glabrous, not maculate. Leaf sheaths with oral setae. Leaf blades lanceolate, 4 to 6 cm long, 5 to 10 mm wide. Lateral veins 2 to 3 pairs. .... (1) *Y. nitakayamemsis*
  2b. Culm sheaths dark brown or stramineous, black maculate, setose, but sometimes glabrous. No oral setae on leaf sheaths. Leaf blades oblong-lanceolate, 5 to 10.5 cm long, 7 to 15 mm wide. Lateral veins 3 to 4 pairs. .... (2) *Y. chungii*
    3a. Culm internodes distinctly longitudinal-elevated-lineate. Young culms pruinose zone under nodes. Culm sheaths glabrous on abaxial surface. Sheath ligules truncate at apex. Primary branches 3 to 7 on nodes. Leaf blades grayish-pubescent beneath. Lateral veins 2 pairs. .... (3) *Y. longissima*
    3b. Culm internodes indistinctly longitudinal-elevated-lineate. Young culms without pruinose zone under nodes. Culm sheaths setose on abaxial surface. Sheath ligules arcuate. Primary branches 10 to 19 on nodes. Leaf blades glabrous beneath. Lateral veins 3 to 4 pairs. ..... (4) *Y. xizangensis*

1. *Yushania nitakayamensis* (Hayata) Keng f.

Culms 45 to 60 cm tall, 4 mm in diameter, with internodes 2 to 3.5 cm long; those grown in shade reaching 2 to 4 m tall, 1 to 2 cm in diameter, with internodes 4 to 10 cm long; erect but arcuate-curved at base; green and glabrous. Nodal ridges slightly swollen. Sheath callus remains. Culm sheaths abscising late, or persistent on unbranched nodes, of thin-papery texture. Oral setae several. Sheath ligules truncate at apex. Sheath blades on upper-culm nodes prominent. Primary branches 1 to 3 (7) on nodes, 5 cm long. Leaf sheaths pilosulose on upper part and rims of abaxial surface, with oral setae. Leaf ligules very short, minute-serrulate.

Distributed in alpine prairies at an altitude of 3000 m in Sichuan, Yunnan, and Taiwan Provinces. Hardy to −15°C.

2. *Yushania chungii* (Keng) Z. P. Wang et F. H. Ye

Culms 2 m tall, 1 cm in diameter. Internodes 15 to 21 cm long, often blackish-dirty. Culm sheath abscising late or persistent, dark brown or stramineous, black-maculate, as long as ½ of internodes, hispid and setulose, sometimes glabrous. Sheath auricles fimbriate. Sheath blades slender. Leaf sheaths ciliate, without oral setae or several setae remain. Leaf auricles rarely remain. Leaf blades oblong-lanceolate. Transverse veinlets evident.

Distributed at an altitude of 2500 to 3800 m in western Sichuan Province. Hardy to −20°C.

3. *Yushania longissima* Yi, sp. nov.

Culms widely separated, 4 m tall, 1 to 2 cm in diameter, with about 21 nodes. Internodes 21 to 34 cm long. Sheath scars not swollen, glabrous. Nodal ridges slightly swollen, glabrous. Intranodes usually blackish-dirty. Leaves 2 to 3 on twigs. Leaf sheaths, purple, pubescent on upper part, not ciliate. No leaf auricles, but grayish setae remain. Leaf ligules purple, truncate at apex, glabrous. Leaf blades papery, lanceolate, 3.5 to 6 cm long, 5 to 8 mm wide; acuminate at apex, rounded or broad-cuneate at base; green and glabrous above, pale green beneath; scabrid at one side of rim only.

Distributed in Xizang (Tibet) Autonomous Region. Hardy to −5°C.

4. *Yushania xizangensis* Yi, sp. nov.

Culms widely separated, 4.5 m tall, 1 to 2 cm in diameter, with about 25 nodes. Internodes 25 to 33 cm long, the longest reaching 40 cm, green and glabrous. Sheath scars slightly swollen to swollen, glabrous or yellowish-brown setose when young. Nodal ridges not swollen or slightly swollen on branching nodes. Primary branches arranged in a half-whorl pattern, nearly upright, without prominent main branch. Leaves (2) 3 (4) on twig. Leaf sheaths pale green, glabrous. Leaf auricles obscure, but gray or grayish-yellow setae evident. Leaf

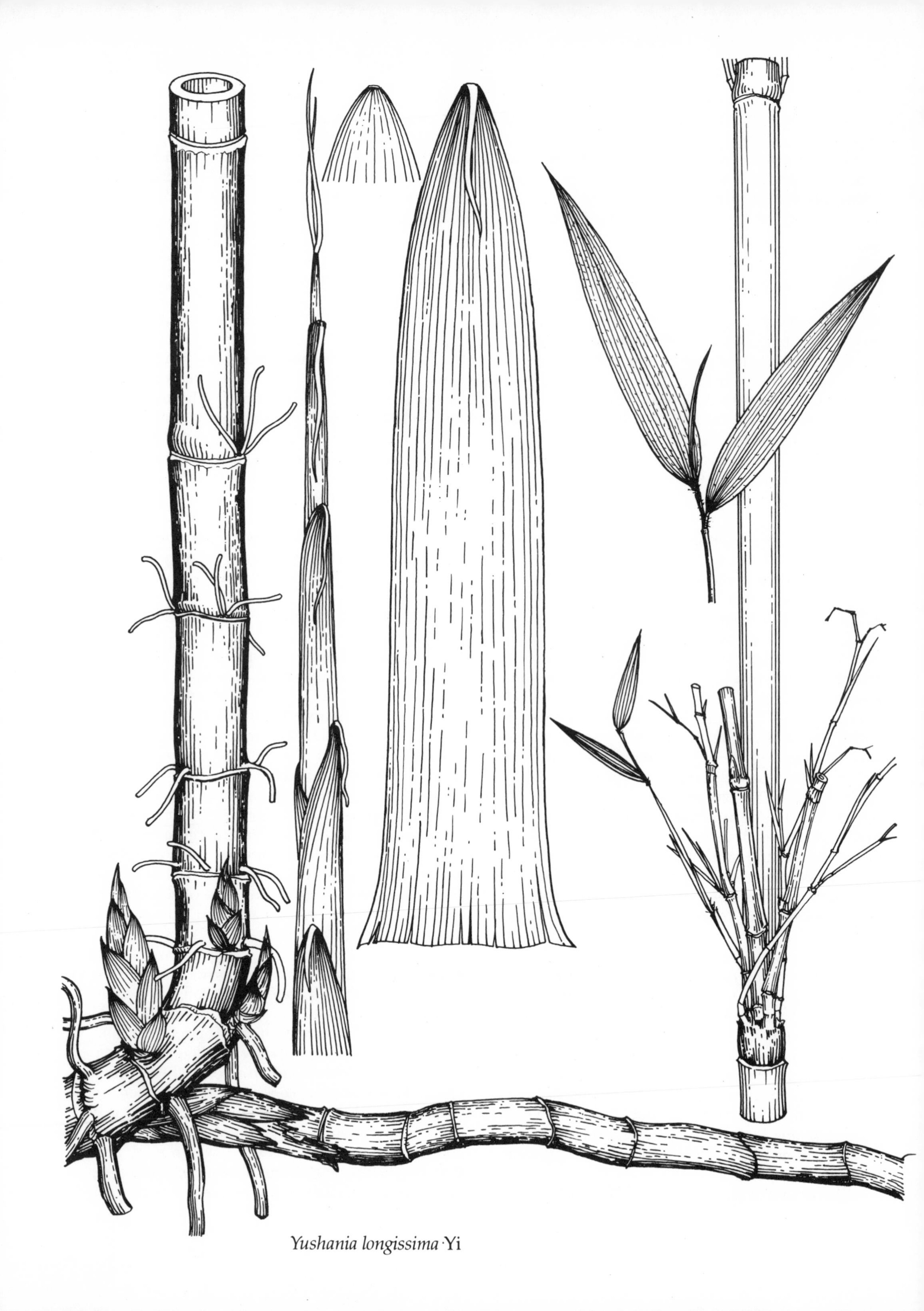

Yushania longissima Yi

ligules truncate or retuse at apex, glabrous. Leaf blades, papery, narrow-lanceolate, (3) 5 to 8 (11) mm long, 5 to 8 mm wide, long-acuminate at apex, cuneate or broad-cuneate; green and sparsely setulose at base above, pale green beneath; serrulate. Transverse veinlets distinct. Leaf stalks compressed, sparsely setulose.

Distributed in Xizang (Tibet) Autonomous Region. Hardy to −5°C.

XX. *Bambusa* Schreber

Tree or shrub-like bamboos, rarely climbing. Rhizomes sympodial. Culms cespitose, cylindrical. Primary branches usually many (rarely a few) clustered on nodes. The main branch usually obviously bigger than the others; occasionally some of secondary branches or twigs replaced by spines. Culm sheaths abscising late or early. Sheath auricles prominent, fimbriate-ciliate. Sheath blades usually erect, and sometimes both erect and reflexed on a few spiny spp. Leaves small, rarely large. Transverse veinlets usually indistinct.

KEY to the Spp. of *Bambusa* Native to China

1a. Branches not spiny. 2
1b. Branches spiny. 20
2a. Shrub-like bamboos, except *Bambusa prasina*. 3
2b. Tree bamboos. 13
3a. Culms abnormal. Nodes closely-set. Internodes short, more swollen at base, pear-form. . . . . (1) *B. ventricosa*
3b. Culms normal. Nodes normally-set. Internodes long, cylindrical. 4
4a. Sheath auricles not obvious or obscure. 5
4b. Sheath auricles prominent. 7
5a. Culm sheaths glabrous. Culms 2 to 7 m tall. Internodes pruinose or brown-setulose when young, pitted after shedding bristles. . . . . (2) *B. multiplex*
5b. Culm sheaths hairy. 6
6a. Culms 7 to 12 m tall. Sheath scars glabrous. Culm sheaths appressed-brown-setose, glabrous on rims. Sheath auricles very small. Leaf blades 18 to 25 cm long, 3 to 5 cm wide. . . . . (3) *B. prasina*
6b. Culms 4 m tall. Sheath callus found on sheath scars. Culm sheaths pale-yellow-hirsute, denser downwards to the base. Sheath auricles absent. Leaf blades 9 to 13 cm long, 9 to 14 mm wide. . . . . (4) *B. strigosa*
7a. Sheath auricles not equal in size. Oral setae absent. 8
7b. Sheath auricles nearly equal in size. Oral setae usually remain. 12
8a. Culm sheaths falcate at apex, both shoulders connected to the falcate apex unequal in length. 9
8b. Culm sheaths truncate at apex, both shoulders connected to the truncate apex nearly equal in length. 11
9a. Sheath ligules serrulate on rim. 10
9b. Sheath ligules nearly entire, 1 mm long. Leaf blades 11 to 16 cm long, 15 to 20 cm wide. . . . . (5) *B. mollis*
10a. Culm nodes slightly swollen. Sheath ligules 0.6 to 1.5 mm long. Leafy blades 7 to 14 cm long, 10 to 14 mm wide. . . . (6) *B. boniopsis*
10b. Culm nodes not swollen. Sheath ligules 2 mm long. Leaf blades 10 to 15 cm long, 13 to 15 mm wide. . . . . (7) *B. contracta*
11a. Lower-culm internodes yellowish-green-longitudinal striate. Sheath ligules 1.5 to 2 mm long. Leaf blades dense-puberulent beneath. . . . . (8) *B. subtruncata*

11b. Lower-culm internodes not striate. Sheath ligules about 1 mm long. Leaf blades scabrid beneath. .... (9) *B. fecunda*

12a. Sheath blades slightly cordate, contracted at base; oral setae many, persistent. .... (10) *B. piscaporum*

12b. Sheath blades prominently cordate-contracted at base; few oral setae, fugacious. .... (11) *B. mutabilis*

13a. Culm internodes bright green, striate. .... (12) *B. vulgaris*

13b. Culm internodes green, not striate or white-striate. 14

14a. Branching low. 15

14b. Branching high. 18

15a. Culm internodes white-striate, at least on the internodes near culm base. 16

15b. Culm internodes not white-striate. 17

16a. Internodes and culm sheaths dense-setulose. .... (13) *B. pervariabilis*

16b. Internodes and culm sheaths glabrous. ...... (14) *B. tuldoides*

17a. Culms 12 m tall. Leaf blades lanceolate or narrow-lanceolate, nearly rounded at base, pilosulose beneath. .... (15) *B. breviflora*

17b. Culms 18 m tall. Leaf blades lanceolate, cuneate at base, white-puberulent beneath. .... (16) *B. lixin*

18a. Culm sheaths appressed-pubescent on abaxial surface when young. Sheath scars oblique. Mid-internodes usually pruinose and acropetal-setose. .... (17) *B. textilis*

18b. Culm sheaths setulose on abaxial surface when young. 19

19a. Culm sheaths of thick-leathery texture, densely blackish-purple setulose on abaxial surface when young. .... (18) *B. dolichoclada*

19b. Culm sheaths of thick-papery texture, appressed-brown-setulose along inner margin of abaxial surface only when young. .... (19) *B. rigida*

20a. Culm sheaths indumentum on abaxial surface. 21

20b. Culm sheaths glabrous on abaxial surface. 31

21a. Culm sheaths indumentum wholly on abaxial surface. 22

21b. Culm sheaths indumentum only on the base of abaxial surface. 27

22a. Culm sheaths densely dark-purple-velvety on both surfaces. .... (20) *B. stenostachya*

22b. Culm sheaths brown-setose on abaxial surface. 23

23a. The two corners (shoulders) asymmetrical on the top of sheath proper. 24

23b. The two corners symmetrical on the top of sheath proper. 26

24a. Culm sheath tops broad, asymmetrical-rounded or nearly truncate. Sheath proper sparsely-appressed-blackish-brown-setose on the abaxial surface, fugacious. The two coupled sheath auricles unequal in size, sometimes obscure. .... (21) *B. gibba*

24b. Culm sheath tops asymmetrical-broad-arcuate. 25

25a. Culm sheaths longitudinal-striate and indistinctly setose on abaxial surface. The two coupled sheath auricles unequal in size. Lower-culm internodes sometimes white striate, glabrous. Leaf blades sparsely-minute-hairy. .... (22) *B. dissemulator*

25b. Culm sheaths appressed-dark-brown-setose on abaxial surface. The two coupled sheath auricles markedly unequal in size. Culms appressed-brown-setulose, sparsely or densely when young. Leaf blade densely pubescent beneath. .... (23) *B. diaoluoshanensis*

26a. Culm sheaths usually acropetally-appressed-dark-brown-short-setose on abaxial surface. Sheath proper tops acuminate. Sheath auricles indistinct. Culm internodes usually slightly curving. Nodes and zones nearly under nodes usually appressed-rusty-tomentose. .... (24) *B. bambos*

26b. Culm sheaths dark-brown-setose on abaxial surface. Sheath proper tops nearly truncate. The two coupled sheath auricles unequal in size. The 1st to 3rd internodes from culm base up, usually purple-maculate, girdled-brown-sericeous on intranodes and under nodes. .... (25) *B. insularis*

27a. The two coupled sheath auricles unequal in size. 28

27b. The two coupled sheath auricles nearly equal in size, setulose. Culm internodes glabrous. Sheath proper brown setose only on the base of abaxial surface. Sheath ligules serrulate on rim, grayish-white-hispid on both surfaces. .... (26) *B. sinospinosa*

28a. Leaf blades glabrous on both surfaces. Lower-culm internodes yellowish-brown-setulose, pitted after shedding bristles. Sheath proper tops asymmetrically truncate. Upper sheath proper reddish-brown-striate; grayish setulose on base. Sheath auricles prominent. .... (27) *B. funghomii*

28b. Leaf blades dense-pubescent beneath. 29

29a. A few nodes near culm base grayish-sericeous on intranodes and under nodes, as girdles. 30

29b. A few intranodes near culm base pale-brown-sericeous and girdled with aerial roots. Sheath proper tops broad-arcuate, and triangular on one corner only; dark-brown setose near the suture on the base of abaxial surface. .... (28) *B. prominens*

30a. Sheath proper tops slightly asymmetrically-arcuate, glabrous on the abaxial surface or setose only on the base; yellow-longitudinal-striate on the outer rims. .... (29) *B. indigena*

30b. Sheath proper tops asymmetrically-broad-arcuate, appressed-dark-brown-setose only on the middle part of the base of abaxial surface. .... (30) *B. xiashanensis*

31a. Leaf blades lanceolate, nearly glabrous or minute-hairy. 32

31b. Leaf blades slender-lanceolate, rarely linear (in case of *B. subaegualia*). 33

32a. Culm internodes glabrous. Sheath proper tops broad and asymmetrically truncate; elevated-striate on the abaxial surface. Sheath auricles oblong. The two coupled auricles short-hirsute on both surfaces, markedly unequal in size. Sheath ligules 3 to 4 mm long. .... (31) *B. malingensis*

32b. A few culm nodes near base grayish-white- or pale-brown-sericeous on intranodes and also girdled under nodes. Sheath proper tops asymmetrically triangular. Sheath auricles narrow-linear and very long-decurrent. Sheath ligules 1 to 2 mm long. .... (32) *B. angustissima*

33a. Sheath proper tops truncate or slanted-truncate. 34

33b. Sheath proper tops asymmetrically broad-arcuate or asymmetrically broad-triangular. 36

34a. Leaf blades pubescent beneath. Sheath ligules minute-serrulate, 3 mm long. 35

34b. Leaf blades glabrous on both surfaces. Sheath ligules entire or nearly entire, about 4 mm long, densely white-setulose on abaxial surface. A few internodes on lower-culm sometimes purple- or pale-green-longitudinal striate. Sheath proper tops truncate; olive-green, pale-purple-maculate or striate. .... (33) *B. lapidea*

35a. A few internodes from base up markedly shorter, grayish-white-sericeous on intranodes and girdled under nodes. Sheath proper tops nearly truncate and cornute only on one corner. Sheath ligules strigose on abaxial surface. .... (34) *B. corniculata*

35b. Culm internodes glabrous, but a few nodes from the culm base up girdled-grayish-white-sericeous under nodes. Sheath proper tops slanted-truncate. Sheath ligules ciliolate or nearly glabrous. .... (35) *B. ramispinosa*

36a. Leaf blades pubescent beneath. 37

36b. Leaf blades setulose or glabrescent beneath. A few internodes from culm base up dense-basipetal-setose. Lower-culm internodes reddish-brown-setose. Sheath proper tops convex or concave, glabrous or glabrescent. Sheath auricles very prominent. Leaf blades slender-lanceolate. .... (36) *B. rutila*

37a. Culm internodes glabrous. A few nodes near culm base girdled-grayish-white-sericeous. Sheath proper tops asymmetrically broad-arcuate. The two coupled sheath auricles unequal in size, strongly rugose. Leaf blades slender-lanceolate, smaller. .... (37) *B. macrotis*

37b. Culm internodes and nodes glabrous. Sheath proper tops slightly asymmetrically broad-triangular or nearly broad-arcuate, usually with 1 to 2 stripes (pale-yellowish-green in color) on outer rim. Sheath auricles indistinct. Leaf blades linear. .... (38) *B. subaequalia*

1. *Bambusa ventricosa* McClure

Shrub-like bamboos. The length and diameter of culms vary with the cultural conditions. Normal culms 2.5 to 5 m tall, 1.2 to 5.5 cm in diameter, internodes cylindrical, 10 to 20 cm long. Abnormal culms 25 to 50 cm tall, 2.5 cm in diameter, internodes shortened, with swollen base, pear-formed. Young culms dark green. Old culms olive-yellow. Primary branches 1 to 3 on nodes. Leaves 7 to 13 on twigs. Leaf blades ovate-lanceolate to oblong-lanceolate, 12 to 21 cm long, 13 to 33 mm wide; pilosulose beneath. Lateral veins 5 to 9 pairs.

Native to Guangdong Province. Grown in gardens in southern China, and as pot plants in the north. Tender.

2. *Bambusa multiplex* (Lour.) Raeusch.

Culms 2 to 7 m tall, 5 to 22 mm in diameter. Internodes green, pruinose or brown-setulose when young, pitted after shedding bristles. Primary branches many clustered on nodes, the main branch slightly thicker than the others. Leaves usually 5 to 10 on twigs, nearly distichous, linear-lanceolate, 4 to 14 cm long, 6 to 12 mm wide; green and glabrous above; powdery-white or pale-green, indumentum and scabrid beneath. Lateral veins 4 to 8 pairs, no transverse veinlets.

Distributed from Yangtze River region to southern China and Taiwan Province. One of the hardiest clumped bamboos, tolerates −15°C in sheltered places. Ornamental and pot-grown as penjing.

Bambusa multiplex var. *nana* (Roxb.) Keng f.

Culms 1.5 to 2 m tall, 0.5 to 10 mm in diameter. Smaller than the sp. Usually branching from the 2nd node to the culm base. Primary branches many on nodes in semi-whorl pattern. Leafy twig single on primary branch nodes. Leaves more than 10 on twigs, nearly distichous. New leaves continuously

Bambusa multiplex (Lour.) Raeusch.

grown on twig tip while old leaves on lower twig abscising. Leave blades small, 1.7 to 5 cm long, 3 to 8 mm wide. Bamboo shoots appear in summer and autumn.

Distributed in Yangtze River Provinces. Ornamental and pot-grown as penjing.

Bambusa multiplex var. *luta* Wen, var. nov.

Culms 3 m tall, 0.8 cm in diameter. Young culms green and puberulent, soon turning yellow and glabrous. Sheath blades glabrous. Sheath auricles prominently protruding, falciform, clasping. Leaf blades smaller than the spp.

Cultivated plant. Hardy to −6°C.

Bambusa multiplex f. *alphonsokarri* (Mitf.) Sasaki

Culms slender, 10 m tall, 1 to 2.5 cm in diameter. The 1st internodes from base comparatively long, about 20 cm. The longest internodes on mid-culm reaching to 50 cm long. Upper part of internodes scattered-acropetal-setulose, pitted after shedding bristles. Culms golden yellow. A few internodes near base 1-to-10-green-longitudinal striate, stripes of uneven width.

Cultivated plant. Hardy to −10°C.

3. *Bambusa prasina* Wen, sp. nov.

Culms 7 to 12 m tall, 6 to 9 cm in diameter. Internodes green, 35 cm long. Young culms hirsute, becoming sparse when old. No pruinose zone under nodes. Sheath scars glabrous. Branching high. Culm sheath green, soon turning to yellow. Leaves 6 to 8 on twigs. Leaf sheaths with longitudinal veins, setose, soon becoming glabrous. Leaf auricles falciform, ciliolate. Leaf ligules truncate, scabrous. Leaf blades long-elliptic; truncate or rounded at base, acute at apex and excurrent; glabrous on both surfaces. Lateral veins 7 to 10 pairs, tertiary veins 9 to 10 between two lateral veins. Transverse veinlets indistinct.

Distributed in Zhejiang and Fujian Provinces. Tender.

4. *Bambusa strigosa* Wen, sp. nov.

Culms 4 m tall, 1 cm in diameter. Internodes green, white-pubescent; much denser when young. Nodal ridges swollen, girdled sheath callus exserted on sheath scars. Leaves 7 to 11 on slender twigs. Leaf sheath puberulent. Leaf auricles moderately prominent, ciliolate. Leaf ligules not prominent. Leaf blades narrow-lanceolate, 9 to 13 cm long, 9 to 14 mm wide; green and glabrous above, powdery-green and pubescent beneath; truncate at base, acute and excurrent at apex. Lateral veins 3 to 6 pairs, transverse veinlets none.

Distributed in Xunwu Prefecture of Jiangxi Province. Hardy to −10°C.

5. *Bambusa mollis* Chia et H. L. Fung, sp. nov.

Culms 5 to 8 m tall, 2.5 to 3.5 cm in diameter. Internodes 35 to 40 cm long. The 1st to 2nd nodes on culm base densely grayish-white-sericeous on intranodes. Branching usually from the 3rd to 5th node from culm base. Primary branches many, clustered on nodes, of even thickness. Sheath proper tops asymmetrically arcuate, glabrous on abaxial surface. Sheath blade base as broad as 3/7 of sheath proper top. The two coupled sheath auricles unequal in size, oblong-lanceolate. Sheath ligules glabrous. Leaf blades pubescent beneath.

Distributed in Guangxi Province. Tender.

6. *Bambusa boniopsis* McClure

Culms 2 to 4.5 m tall, 1 to 2 cm in diameter. Culm wall thick. Internodes cylindrical, slightly curving, glabrous, slightly farinose when young; green, turning to somewhat yellowish when old, but not green-striate. Nodes slightly swollen, glabrous. Leaf blades linear-lanceolate; acuminate and subulaliferous at apex; asymmetrical at base; pilose or glabrescent above; whitish-green and dense-pubescent beneath. Leaf sheath glabrous. Leaf auricles prominently decurrent, ciliate.

Distributed in Hainan Island of Guangdong Province. Tender.

7. *Bambusa contracta* Chia et H. L. Fung, sp. nov.

Culms 5 to 6 m tall, 2 to 3 cm in diameter. Internodes 40 to 57 cm long. Nodes not swollen and glabrous. Branching usually from the 4th to 6th node up from the culm base. Primary branches many, clustered on nodes, of nearly even thickness. Sheath proper tops asymmetrically arcuate, and decurrent on one side only, sparsely pruinose on abaxial surface. Sheath blade slightly rounded or nearly cordate at base, as broad as ¼ of sheath proper top. The two coupled sheath auricles unequal in size, wavy-rugose, oblong to lanceolate. Sheath ligules 2 mm long, sparse-serrulate. Leaf blades dense-pubescent beneath.

Distributed in Guangdong Province. Tender.

8. *Bambusa subtruncata* Chia et H. L. Fung, sp. nov.

Culms 4 to 5 m tall, 2 to 2.5 cm in diameter. Internodes 25 to 30 cm long, glabrous. A few internodes near culm base yellowish-green-striate. The 1st to 3rd nodes from culm base girdled-grayish-white sericeous on intranodes and under nodes. Branching from the 3rd or 4th node up from culm base. The central primary branch on each node prominently thicker and longer than the others. Culm sheath yellowish-green-longitudinal-striate; truncate at apex. The base of sheath blade as broad as 3/5 of the top of sheath proper. Sheath auricles broad-elliptic to elliptic. The two coupled auricles unequal in size. Leaf blades 8 to 15 cm long, 0.9 to 1.3 cm wide, dense-pubescent beneath.

Distributed in Guangdong Province. Tender.

9. *Bambusa fecunda* McClure

Culms 3 to 5 m tall, 1 to 1.5 cm in diameter. Internodes cylindrical, slightly curving, glabrous. Grayish-white sericeous under nodes. Nodes slightly swollen. Culm sheaths glabrous and waxy. Sheath blades erect, longer than sheath proper. Sheath proper tops truncate. The two coupled sheath auricles unequal in size, the larger one oblong. Sheath ligules nearly entire or serrulate. Primary branches many, clustered on nodes, of even size. Leaf blades linear-lanceolate to lanceolate, 9 to 16 cm long, 1 to 1.4 cm wide; acuminate and scabrous-subulate at apex; broad cuneate or nearly truncate at base; scabrid beneath. Leaf sheath elevated-striate, glabrous. Leaf auricles decurrent, with a few oral setae.

Distributed in Hainan Island of Guangdong Province. Tender.

10. *Bambusa piscaporum* McClure

Culms 5 to 7 m tall, 2.5 to 3.6 cm in diameter. Internodes long, glabrous, turning to yellowish when old, but not green-striate. Nodes not swollen. Culm sheath abscising early. Sheath blades erect, lanceolate-triangular, long and acuminate at apex, slightly cordate-contracted at base; glabrescent on abaxial surface; scabrid and appressed-short-hirsute on adaxial surface. Sheath proper tops broad-falcate, glabrous or fugacious-appressed-hirsute on abaxial surface. Sheath auricles small, nearly equal in size. Branches slender, glabrous. Leaf blade linear-lanceolate, 6 to 22 cm long, 1 to 2.2 cm wide; acuminate at apex, obtuse at base; dense-pubescent beneath. Leaf auricles oblong or lanceolate, fugacious, with 4 to 8 setae on rims; oral setae a few, fugacious.

Distributed in Hainan Island of Guangdong Province. Tender.

11. *Bambusa mutabilis* McClure

Culms 4 to 5 m tall, 3.1 cm in diameter. Internodes long, glabrous or sparsely appressed-setose; pruinose when young, green. Internodes on culm base usually pale-purple-elongated-maculate. Nodal ridges slightly swollen. Sheath blades erect, broad-lanceolate, acuminate at apex; glabrous on abaxial surface; acropetal-setose between veins on adaxial surface. Sheath proper tops truncate, nearly flat-striate on abaxial surface, lucid, glabrous. Branches slender. Branch sheath densely short-hispid. Leaf blades lanceolate-linear, 7 to 23 cm long, acuminate and subulate at apex, obtuse or broad-cuneate at base; dense pubescent beneath. Leaf sheaths ribbed, the upper part pilosulose. Leaf auricles prominent, sometimes ciliate. Leaf ligules short, puberulent on abaxial surface, truncate at apex.

Distributed in Hainan Island of Guangdong Province. Tender.

12. *Bambusa vulgaris* Schrad

Culms 20 m tall, 15 cm in diameter. The tops of old culms arching. Primary branches many on nodes. Sheath proper tops very broad. The base of sheath blades much narrower. Leaf blades 25 cm long, 3.2 cm wide. Rims and the under surface of leaves scabrous. Transverse veinlets none.

Cultivated plant used in the landscape. Tender.

Bambusa vulgaris var. *striata* Gamble

Culms 15 m tall, 5 to 6 cm in diameter. Internodes 20 to 25 cm long, bright yellow, distinct-green striate. Nodes swollen. Leaf blades linear-lanceolate, 20 cm long, 1.5 to 4 cm wide, long-acute at apex, narrowed at base to stalk.

Distributed in Guangdong and Guangxi Provinces. Highly ornamental. Tender.

13. *Bambusa pervariabilis* McClure

Culms erect or nearly erect, 15 m tall, 6 cm in diameter. Culm tops not drooping. Branching low. Internodes densely setulose. Primary branches spreading horizontally. Branch and twig sheaths purple when young. Culm sheaths variable in shape and indumentum.

Distributed in Guangxi and Guangdong Provinces. Tender.

14. *Bambusa tuldoides* Munro

This sp. differs from *B. pervariabilis* McClure on two points mainly, i.e. culm internodes and culm sheaths glabrous; the number of primary branches on nodes greater than *B. pervariabilis.*

Distributed in Guangdong and Taiwan Provinces. Tender.

15. *Bambusa breviflora* Munro

Culms erect, strong, 12 m tall, 5 to 8 cm in diameter. White zone above nodes. Branching low. Culm wall thick. Primary branches many, clustered on nodes. Leaf blades 4 to 15 cm long, 12 to 18 mm wide.

Distributed in Guangdong, Guangxi, and Taiwan Provinces. Tender.

16. *Bambusa lixin* Hsueh et Yi, sp. nov.

Culms 15 to 18 m tall, 5 to 8 cm in diameter. Internodes 38 to 42 cm long, green, glabrous, pruinose when young. Sheath scars swollen, glabrous, sheath callus exserted. Nodal ridges not swollen. Intranodes girdled-white-lucid-tomentose. Primary branches many on nodes, one (rarely a few) among them extra-strong, slightly pruinose under nodes. Culm sheaths densely brownish-black or yellowish-brown-setose. Sheath auricles prominent, glabrous except for yellowish-brown-ciliate. Sheath ligules triangular or ligulate-protruding, dark-greenish-brown, glabrous, densely brownish-yellow-sericeous-ciliate. Sheath blades ovate-triangular, green, slightly pruinose, glabrous; serrulate on rim of one side, nearly smooth on rim of the other side. Leaves 5 to 7 on twigs. Leaf sheath green or yellowish-green, glabrous. Leaf auricles obscure, oral setae none. Leaf ligules truncate, pale-yellow-green, glabrous. Leaf blade 9 to 25 cm long, 1.5 to 2.8 cm wide, green and glabrous above, grayish-green beneath; acuminate at apex, cuneate at base; serrulate and scabrid on rim. Lateral veins 5 to 10 pairs. Transverse veinlets indistinct.

Distributed at an altitude of 900 to 1100 m in Xizang (Tibet) Autonomous Region. Hardy to −10°C.

17. *Bambusa textilis* McClure

Culms 8 to 12 m tall, 5 to 6 cm in diameter. Culm tops slightly drooping. Internodes cylindrical, 35 to 50 cm long; pruinose when young, glochidiate. Sheath scar slanted. Culm sheath deciduous, pubescent when young. Sheath blades long-triangular, as long as or a little less than sheath proper. Sheath auricles small, long-elliptic. Sheath ligules slightly arcuate. Branching high. Primary branches small and short, 10 to 12 clustered on nodes, 2 to 3 branches a little longer than the others. Leaves 8 to 14 on twigs. Leaf blades 10 to 25 cm

long, 1.5 to 2.5 cm wide. Lateral veins 5 to 6 pairs.

Distributed in Guangdong, Guangxi, Fujian, and southern Yunnan. Tolerates a minimum temperature of −3°C. Ornamental.

Bambusa textilis var. *gracilis* McClure

Culms more slender than the sp. Internodes scattered-setose or glabrescent, muricate.

Distributed in Guangdong Province. Ornamental or hedge plant.

Bambusa textilis var. *albostriata* McClure

Internodes white-striate. Ornamental.

Bambusa textilis var. *maculata* McClure

Internodes maculate, and sometimes purplish-lineate. Ornamental.

Bambusa textilis var. *fusca* McClure

Culms dark-brownish. Ornamental.

Bambusa textilis var. *glabra* McClure

Culms yellowish, glabrous. Ornamental.

18. *Bambusa dolichoclada* Hayata

Culms 15 to 20 m tall, 5 to 13 cm in diameter. Internodes 25 to 55 cm long, densely glaucous when young and turning green later. Old culms yellowish-red or brownish. Nodes slightly swollen. Sheath scars blackish-brown puberulent. Primary branches and twigs many, clustered on nodes and branch nodes. Spinescent buds on branch nodes. Leaves 8 to 14 on twigs. Leaf blades narrow-lanceolate to broad-lanceolate, 12 to 20 cm long, 14 to 22 mm wide; serrulate; glabrous above, scattered puberulent beneath when young, then glabrous. Lateral veins 8 to 9 pairs, tertiary veins 9 to 10 between lateral veins.

Distributed in Fujian and Taiwan Provinces. Tender.

In their writings and poems, Chinese intellectuals of the past often expressed their envy of the simple life of the fisherman. Hence, in Chinese paintings, water scenery with fishing boats is especially favored.

19. *Bambusa rigida* Keng et Keng f.

Culms 5 to 12 m tall, 2 to 6 cm in diameter, erect, or slightly arching at top. Internodes glabrous, glaucous especially dense on the part surrounded by sheath proper. Internodes on lower-culm 23 to 35 cm long, on upper culm 45 cm long. Culm sheaths grayish-brown, as long as ½ of internodes, persistent. Primary branches several to 15 on nodes, fastigiate or spreading. Twigs 2 to several on branch nodes. Leaves 5 to 12 on upper part of twigs. Leaf blades oblong, or narrow-oblong, 8 to 24 cm long, 9 to 27 mm wide; dark green and glabrous above, grayish-green and pilosulose beneath, serrulate on both sides or on one side only. Lateral veins 4 to 9 pairs.

Distributed in Guangdong, Guangxi, and Sichuan Provinces. Tender.

20. *Bambusa stenostachya* Hackel

Culms big, 15 to 20 m tall, 8 to 15 cm in diameter. Internodes near culm base almost solid, pruinose when young, 25 to 40 cm long. Sheath scars more swollen than nodal ridges. Culm sheath large, tenacious,. greenish-yellow, densely dark-purple velvety. Sheath auricles prominent. Sheath blades ovate-triangular. Branching low. Primary branch usually single (rarely 2 to 3) on lower-culm nodes. Spines 1 to 5 on nodes of small branches. Leaves 5 to 12 on twigs. Leaf sheaths glabrous. Leaf auricles exserted. Leaf ligules truncate. Leaf blades slender lanceolate, 6 to 18 cm long, 11 to 20 mm wide, acute at apex, cilia-like serrulate, mid-rib elevated on under surface. Lateral veins 4 to 6 pairs.

Distributed in Guangdong, Guangxi, and Taiwan Provinces. Tender.

21. *Bambusa gibba* McClure

Culms 3 to 8 m tall, 2.5 to 5 cm in diameter. Internodes pruinose, comparatively densely appressed-acropetal-white-setose, fugacious later. Sheath blades erect, fugacious, narrow-triangular to triangular, glabrous on abaxial surface, scabrous on adaxial surface. Secondary branches on lower-culm nodes sometimes deformed to weak spines.

Distributed in Guangdong and Guangxi Provinces. Tender.

22. *Bambusa dissemulator* McClure

Culms erect or semi-erect, somewhat arching at top, 15 m tall, 7.5 cm in diameter. Internodes nearly cylindrical, green; white-striate on lower part when young; glabrous. Nodal ridges slightly swollen. Primary branch single on lower-culm nodes, with 2 secondary branches and 2 to 6 slender spines at base. The secondary branches sometimes bearing a few spines. Twigs spineless. Leaves 5 to 14 on twigs. Leaf blades lanceolate, 7 to 17 cm long, 10 to 15 mm wide. Lateral veins 3 to 6 pairs, indistinct. Transverse veinlets none.

Distributed in Guangdong and Guangxi Provinces. Tender. Cultivated around villages.

Bambusa dissemulator var. *albinodia* McClure

Culms base nodes white-circular-striate.

Bambusa dissemulator var. *hispida* McClure

Nodes, internodes and culm sheaths obviously setulose.

23. *Bambusa diaoluoshanensis* Chia et H. L. Fung, sp. nov.

Culms 10 m tall, 4 to 5 cm in diameter. Lower-culm nodes girdled with pale-brown bristles and glaucous under nodes. The 1st to 3rd nodes from culm base grayish-sericeous. Branching from the 1st node at culm base. Secondary branches on lower-culm nodes sometimes deformed to weak spines. Leaf blades 7.5 to 16 cm long, 1.2 to 1.8 cm wide.

Distributed in Guangdong Province. Tender.

24. *Bambusa bambos* (Linn.) Voss

Culms 4 to 15 m tall, 2.5 to 8 cm in diameter. Internodes green, 24 cm long. Leaf blades lanceolate, 4 to 10 cm long, 7 to 11 mm wide, acuminate at apex, rotund or sometimes nearly truncate at base; pilose or glabrescent. Leaf sheaths glabrous, striate. Leaf auricles prominent, fimbriate on rim. Leaf ligules prominent, fimbriate on rim. Primary branches many clustered on nodes, the central one longer and thicker, reaching 1.5 m long. Leafless twigs deformed to spines.

Distributed in southern China. Tender.

25. *Bambusa insularis* Chia et H. L. Fung, sp. nov.

Culms 8 to 10 m tall, 4 to 5 cm in diameter. Branching from the 1st to 2nd nodes at culm base. Secondary branches on lower-culm nodes often deformed to weak spines. Leaf blades 8 to 14 cm long, 1.1 to 1.5 cm wide; pilose near base on upper surface, dense pubescent beneath.

Distributed in Guangdong Province. Tender.

26. *Bambusa sinospinosa* McClure

Giant bamboo, culms 10 to 24 m tall, 5 to 15 cm in diameter. Internodes green, glabrous. Sheath scars densely brown-setose. Primary branches 3 on nodes. Spines 2 to 3, in 'T' pattern, on every branch node. Leaves 6 to 8 on twigs. Leaf blades slender-lanceolate, 6 to 20 cm long, 6 to 20 mm wide. Lateral veins 4 to 7 pairs. Bamboo shoots appear from May to August.

Widely distributed in Guangdong, Guangxi, Sichuan, Guizhou, and southeastern Yunnan Provinces. Tender.

27. ***Bambusa funghomii*** McClure

Culms erect or semi-erect, upright or arching at top, 15 m tall, 6.5 cm in diameter. Nodes swollen. Lower-culm nodes gesticulate, glabrous. Internodes green, broad-grooved on branching side. Branching low. Primary branches 3 to many on nodes, except a few nodes near culm base bearing 1 or 2 branches. The central primary branch bigger than the others, sometimes up to 5 m long on lower-culm nodes. Branches thorny. The secondary branches on lower-central branches deformed to thorns; but those on upper part thornless, bearing 3 to 6 leaves instead. Leaf blades slender-lanceolate, 5 to 15 cm long, 5 to 25 mm wide, glabrous. Lateral veins 3 to 6 pairs. Transverse veinlets on new leaves prominent.

Distributed in Guangdong and Guangxi Provinces. Tender.

28. ***Bambusa prominens*** H. L. Fung et C. Y. Sia, sp. nov.

Culms 10 to 15 m tall, 5 to 7 cm in diameter. Branching from the 1st node at base. Secondary branches on lower-culm nodes sometimes deformed to weak spines. Leaf blades 5 to 25 cm long, 2 to 2.5 cm wide, densely pubescent beneath.

Distributed in Sichuan Province. Rapidly growing plant. Tender.

29. ***Bambusa indigena*** Chia et H. L. Fung, sp. nov.

Culms 10 to 14 m tall, 4.5 to 7 cm in diameter, sparse appressed-setulose when young. Branching from the 1st to 2nd nodes at base. Secondary branches on lower-culm nodes often deformed to weak spines. Leaf blades 6.5 to 12 cm long, 1.3 to 1.7 cm wide.

Distributed in Guangdong Province. Rapidly growing. Tolerates draught. Commonly used for hedges. Tender.

30. ***Bambusa xiashanensis*** Chia et H. L. Fung, sp. nov.

Culms 12 to 13 m tall, 4.5 to 5.5 cm wide. Branching from the 1st or 2nd nodes at culm base. Secondary branches on lower-culm nodes sometimes deformed to weak spines. Leaf blades 10 to 20 cm long, 1.5 to 2 cm wide.

Distributed in Guangdong Province. Tender.

31. ***Bambusa malingensis*** McClure

Culms erect or semi-erect, 10.5 m tall, 5.9 cm in diameter. Sheath scars slightly swollen. Small branches on lower-culm nodes sometimes deformed to small spines. Leaf blades lanceolate, 5 to 17 cm long, 1 to 2 cm wide, acuminate and subulate at apex, broad-cuneate or nearly rounded at base; scabrid along base of midrib above, glabrescent or pilosulose beneath, scabrous on rim. Leaf sheaths elevated-striate, glabrous. Leaf auricles prominent, ciliate, cilia 5 mm long. Leaf ligules short, truncate at apex.

Distributed in Hainan Island of Guangdong Province. Tender.

32. ***Bambusa angustissima*** Chia et H. L. Fung, sp. nov.

Culms 9 m tall, 5 cm in diameter. Culm wall thick. Often branching from culm base. Small branches on lower-culm primary branches often deformed to weak or strong spines. Leaf blades 6 to 9 cm long, 1.1 to 1.5 cm wide; glabrescent beneath.

Distributed in Guangdong Province. Tender.

33. ***Bambusa lapidea*** McClure

Culms erect, 17 m tall, 6.5 cm in diameter. Nodes swollen, glabrous. Internodes green glabrous. Primary branches strong, those near culm base somewhat arching, 3 to 7 on every culm node. Secondary branches often thorny, those on upper-culm nodes usually spineless. Leaves 3 to 12 on twigs. Leaf blades slender-lanceolate, 6 to 29 cm long, 5 to 32 mm wide. Lateral veins 4 to 10 pairs.

Distributed in Guangdong and Gaungxi Provinces. Tender.

34. ***Bambusa corniculata*** Chia et H. L. Fung, sp. nov.

Culms 8 m tall, 4 to 7 cm in diameter. Culm wall 8 mm thick. Often branching at the 2nd to 3rd nodes from culm base. Primary branches 3 to many on nodes, except singly on culm base nodes. Secondary branches on lower-culm nodes often deformed to curved weak spines. Leaf blades 13 to 20 cm long, 1 to 2 cm wide, pubescent beneath.

Distributed in Guangdong and Guangxi Provinces. Tender.

35. ***Bambusa ramispinosa*** Chia et H. L. Fung, sp. nov.

Culms 8 m tall, 3.8 cm in diameter. Lower culm internodes 30 cm long. Branching from the 3rd nodes at culm base. Main primary branches markedly stronger and longer than the others, and swollen at base. Secondary branches on lower-culm nodes sometimes deformed to weak spines. Leaf blades 9.5 to 13 cm long, 1.1 to 1.6 cm wide;

puberulent beneath.

Distributed in Guangxi Province. Tender.

36. ***Bambusa rutila*** McClure

Culms erect or semi-erect, 12 m tall, 6 cm in diameter. Nodes prominently swollen, those on lower culm geniculate. All nodes bearing buds, and densely basipetal-setose. Internodes green. Lower culm internodes pitted after shedding bristles. Lower part of internodes compressed or broad-grooved on branching side. Primary branches several on every node, strong and drooping, 4 to 5 m long, slender curved thorns at base. Secondary branches sometimes deformed to spines. Leaves 10 on spineless twigs. Leaf blades 9.5 to 22 (up to 30) cm long, 1.5 to 3 (4.5) cm wide. Lateral veins and transverse veinlets indistinct.

Distributed in Guangdong Province and Hong Kong. Tender.

37. ***Bambusa macrotis*** Chia et H. L. Fung, sp. nov.

Culms 6 to 7 m tall, 6 cm in diameter. Branches on lower-culm nodes often spiny. Leaf blades 5 to 10 cm long, 7 to 9 mm wide, pubescent beneath, midrib thin.

Distributed in Guangdong Province. Used for hedges. Tender.

38. ***Bambusa subaequalia*** H. L. Fung et C. Y. Sia, sp. nov.

Culms 8 to 12 m tall, 4 to 6 cm in diameter. Internodes 40 to 50 cm long. Culm wall 8 to 10 mm thick. Branching from the 1st node at culm base. Secondary branches on lower-culm nodes often deformed to weak spines. Leaf blades linear, 9 to 15 cm long, 1 to 1.3 cm wide, pubescent beneath. Leaf stalk 1 to 1.5 mm long only.

Distributed in Sichuan Province. Tender.

XXI. *Ampelocalamus* Chen, Wen et Sheng, gen. nov.

Rhizomes sympodial. Culms erect. Internodes cylindrical. Culm tops climbing or drooping. Primary branches 2, 3 to many on nodes. Culm sheaths abscising late or persistent, shorter than internodes. Sheath auricles and leaf auricles long-ciliate, cilia radiating. Sheath ligules long-fimbriate-ciliate on rim. Sheath blades linear-lanceolate to lanceolate, reflexed, shorter or equal to the length of sheath proper. Transverse veinlets not distinct.

KEY to the Spp. of *Ampelocalamus* Native to China

1a. Culms and culm sheaths setose. Sheath auricles and leaf auricles long-ciliate, 2 to 2.5 cm long. Leaf blades larger, of thinner texture, 12 to 30 cm long, setulose on both surfaces. Lateral vein distinct. .... (1) *A. actinotrichus*

1b. Culms and culm sheaths pubescent. Sheath auricles and leaf auricles ciliate, 1 cm long. Leaf blades smaller, of thicker texture, 7 to 20 cm long, glabrous, pale grayish-white beneath. Lateral veins not distinct. .... (2) *A. calcareus*

1. *Ampelocalamus actinotrichus* (Merr. et Chun) S. L. Chen, T. H. Wen & G. Y. Sheng

[*Pleioblastus actinotrichus* (Merr. et Chun) Keng f.]

Culms about 1.5 m tall, 3 to 6 mm in diameter. Internodes 14 cm long. Young culms bristly under nodes, glabrate when old but the verruciform base of bristles or scars remains. Culms sheaths abscising late, green, not striate, not distinctly brown-punctate. Sheath auricles fugacious. Leafy branches clustered, 15 to 35 cm long. Leaves 1 to 5 on twigs. Leaf sheaths persistent, of firm texture, glabrous or setulose on upper parts between veins. Leaf auricles luniform, prominent, long-radiating-ciliate on rim. Leaf ligules truncate at apex, hard, long-fimbriate-ciliate on rim. Leaf blades lanceolate, 12 to 30 cm long, 2 to 4 cm wide; acuminate at apex, narrowed at base to a 1 to 6 mm long stalk; pale green above. Lateral veins 3 to 6 pairs.

Distributed in Hainan Island of Guangdong Province at an altitude of 500 m. Ornamental. Tender.

A clump of *Bambusa ventricosa* (Buddha's belly bamboo) in an elevated bed, Guilin.

2. *Ampelocalamus calcareus* C. D. Chu et C. S. Chao, sp. nov.

Culms erect, 1.5 m tall, 4 to 5 mm in diameter. Culm tops drooping. Internodes 8 to 18 cm long, pubescent, glabrate later. Nodes slightly swollen. Primary branches usually 5 to 7 on nodes, clustered, 50 to 100 cm long, about 2 mm in diameter. Culm sheaths persistent, not maculate, densely white-pubescent, glabrate later, densely white-ciliate on rims. Sheath auricles developed, luniform, clasping. Oral setae dark brown. Leaves 2 to 3 on twigs. Leaf sheaths glabrous, smooth, ciliate on rims. Leaf auricles developed, oral setae fugacious. Leaf ligules short, long-white-ciliate at apex. Leaf blades sub-coriaceous, oblong-lanceolate, 7 to 20 cm long, 1.2 to 3 cm wide, glabrous, whitish beneath. Lateral veins 4 to 7 pairs, not distinct.

Distributed in calcareous hilly regions of Guizhou Province at an altitude of 500 m. Tender.

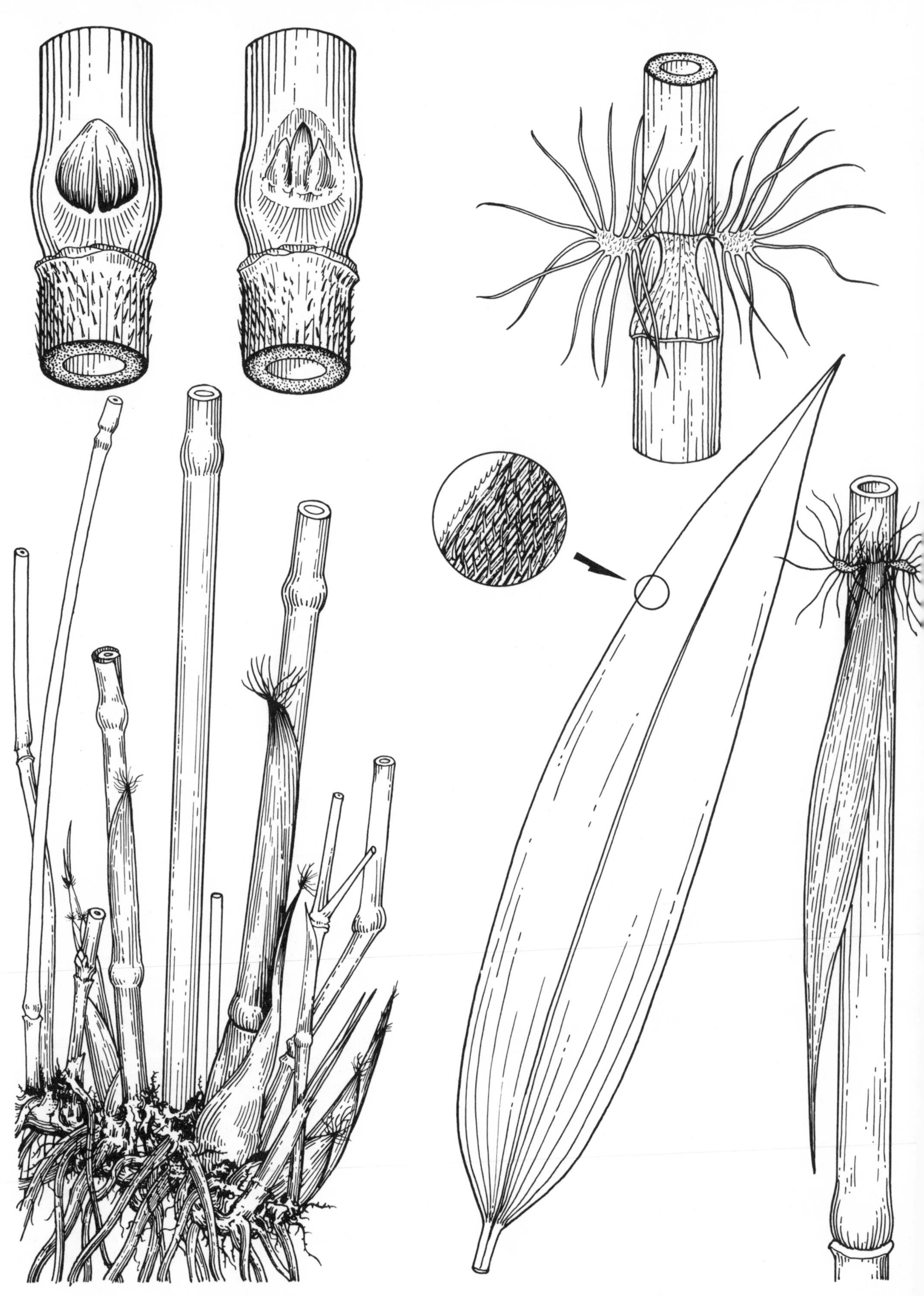

Ampelocalamus actinotrichus (Merr. et Chun) S.L. Chen, T.H. Wen & G.Y. Sheng

XXII. *Lingnania* McClure

Tall or shrub-like bamboos, rhizomes sympodial, culms clumped, usually erect, seldom climbing. Internodes cylindrical, 50 to 100 cm long on midculm. Nodal ridges not swollen. Sheath scars corky and swollen, sometimes barbed. Culm sheaths deciduous; very broad at apex, truncate or arcuate, seldom rounded. Sheath auricles short, broad and level in position. Sheath blades usually reflexed, blade base as broad as only ½ to ¼ of sheath proper apex. Primary branches many on nodes, usually of nearly even thickness. Leaf blades linear-lanceolate to ovate-lanceolate, acuminate at apex, obtuse or cordate at base, usually without transverse veinlets.

KEY to the Spp. of *Lingnania* Native to China

1a. Culms climbing. .... (1) *L. scandens*
1b. Culms erect or nearly erect. 2
2a. Internodes not pruinose. 3
2b. Internodes pruinose. 5
3a. Intranodes downy. Leaf blades larger. 4
3b. Intranodes usually glabrous. Leaf blades smaller, not over 10 cm long and not over 2 cm wide, glabrous. .... (2) *L. remotiflora*
4a. Leaf auricles prominent. Leaf blades 20 cm long, 3 cm wide, glabrous. .... (3) *L. fimbriligulata*
4b. Leaf auricles obsolescent or absent. Leaf blades 16 cm long, 2 cm wide, downy. .... (4) *L. wenchouensis*
5a. Leaf auricles absent. 6
5b. Leaf auricles remain. 7
6a. Culm sheath brown-bristly on abaxial surfaces. Sheath ligules very prominent. Sheath auricles absent. Sheath blades readily reflexed. Lateral veins on leaf blades 7 to 10 pairs. .... (5) *L. farinosa*
6b. Culm sheaths dense-golden-yellow- or brown-bristly on back. Sheath ligules short. Sheath auricles very small. Sheath blades not readily reflexed. Lateral veins on leaf blades 4 to 6 pairs. .... (6) *L. distegia*
7a. The top of culms drooping. Internodes and young culm sheaths dense-pruinose. Sheath blades black. Culm sheaths puberulent on abaxial surfaces. Leaf blades of thin texture, narrower, glabrous on both surfaces, or sparse-hairy on lower surface. .... (7) *L. cerosissima*
7b. The top of culms slightly arching. Internodes and young culm sheaths sparse-pruinose. Sheath blades pale yellowish-green. The abaxial surface of culm sheath densely dark-pubescent (fugacious) on the base only. Leaf blades of thick texture, broader, puberulent beneath at early stage, than glabrate. .... (8) *L. chungii*

1. *Lingnania scandens* McClure

Culms climbing, often reaching 10 m or more. Internodes glabrous. Branches slender. Branch internodes from mid-part up barbed. Leaf blades oblong-lanceolate to linear-lanceolate, 16 to 25 cm long, 2 to 3 cm wide; acuminate with a short awl-shaped apex; glabrous above, puberulent beneath; glabrate; with transverse veinlets. Leaf sheaths thick, markedly striate. Oral setae absent. Leaf ligules concave at apex, ciliate or serrulate. Leaf stalks short and thick, glabrous.

Distributed in Hainan Island of Guangdong Province. Tender.

2. *Lingnania remotiflora* (Kuntz.) McClure
[*Bambusa sponosa* Roxb.]

Culms 10 m tall, 5 cm in diameter. Internodes glabrous. Sheath scars swollen, glabrous or nearly glabrous. Leaf blades lanceolate, 5.5 to 8.5 cm long, 1 to 1.5 cm wide, acuminate, with awl-shaped apex; truncate or obtuse at base; glabrous, or scabrid beneath base. Leaf sheath markedly striate, glabrous on back and margins. Leaf auricles olive-green, short-bristly or glabrous. Leaf ligules very short.

Distributed in southern China. Tender.

3. *Lingnania fimbriligulata* McClure

Culms 6 to 12 m tall, 3.3 to 7.2 cm in diameter. Culm tops erect to slightly arching. Internodes dark green to blackish-green, young culm internodes darker in shade. Sheath scars swollen. All the nodes on culms with visible buds. Primary branches not even in size, the main branches longer and thicker than the others. Leaf blades lanceolate to oblong-lanceolate, 9 to 20 cm long, 1 to 3 cm wide, glabrous. Leaf sheaths usually glabrous, markedly striate. Leaf auricles small, easily broken and crisp, abscising early. Leaf ligules short, fimbriate at apex.

Distributed in Hainan Island of Guangdong Province. Tender.

4. *Lingnania wenchouensis* Wen, sp. nov.

Culms 16 m tall, 8 to 10 cm in diameter. Internodes green, 37 to 50 cm long, pubescent at first, then glabrate. Intranodes downy at first, then glabrate. Culm nodes not prominent, sheath scars swollen. Sheath callus 15 mm wide. Branching low. Leaves 7 to 12 on twigs. Leaf sheath pubescent or glabrous, with prominent longitudinal veins. Leaf auricles ovate, deciduous, ciliate. Leaf ligules obtuse. Leaf stalks 2 mm long, glabrous. Leaf blades broad-lanceolate, serrulate, 9 to 16 cm long, 1.2 to 2 cm wide, acute at apex, base blunt, downy beneath. Lateral veins 6 to 7 pairs, transverse veinlets absent.

Distributed in Zhejiang and Fujian Provinces. Tolerates light frost.

5. *Lingnania farinosa* (Keng et Keng f.) Keng f.
[*Sinocalamus farinonus* Keng et Keng f.]

Culms 10 m tall, slightly arching at top, 4 to 6 cm in diameter. Internodes 10 to 21 cm long, glabrous; green and pruinose when young, brownish yellow and glossy when mature. Sheath scars prominent, with sheath callus; also with golden appressed downy zone below sheath scar and paler appressed downy zone above sheath scars when young, glabrate later. Culm nodal ridges swollen. Primary branches many clustered on nodes, fastigiate on upper culm and spreading on mid-culm (primary branches on top nodes usually in a semi-whorl pattern). Leaves 4 to 9 on twigs. Leaf sheaths glabrous. Leaf ligules truncate at apex. Leaf blades lanceolate, thin, 9 to 17 cm long, 1.5 to 2.5 cm wide, green and glossy above, pale green and white-puberulent beneath. The sizes and form of leaves usually variable.

Distributed in Guangxi, Guizhou and Sichuan Provinces. Tender.

6. *Lingnania distegia* (Kent et Keng f.) Keng f.
[*Sinocalamus distegius* Keng et Keng f.]

Culms 10 m tall, erect, slightly arching at top, 4.5 cm in diameter. Internodes cylindrical, green at first, then yellow, 20 to 50 cm long. Upper part of internodes sparse-pruinose and white-setulose when young. Sheath scars densely brownish-yellow-barbed, glabrate later. Nodal ridges not prominent. Primary branches on top-culm nodes crowded to over ½ of the node. Branch nodal ridges swollen. Leaves over 10 on end of twigs. Leaf sheaths straw-yellow to brown. Leaf blades long-lanceolate, 5 to 16 cm long, 8 to 16 mm wide; dark green and glabrous above, grayish and white-puberulent beneath; serrulate.

Distributed in Sichuan Province. Tolerates light frost.

7. *Lingnania cerosissima* (McClure) McClure

Culms 15 m tall, 5 cm in diameter, tops drooping. Internodes 60 cm long, glabrous. Culm sheaths much shorter than internodes, firm, abscising late, a wide zone of corky callus remains after shedding of sheath. Sheath auricles long and narrow. Sheath blades blackish, ovate-lanceolate, strongly reflexed, bristly near base. Leaves 4 to 8 on twigs. Leaf blades slender-lanceolate, 20 cm long, 2 cm wide, acuminate at apex; scabrous along midrib above, glabrous or puberulent beneath. Lateral veins 5 to 7 pairs.

Distributed in southern China. Tender.

8. *Lingnania chungii* (McClure) McClure

Culms 3 to 10 m tall, 5 cm in diameter. Internodes pale yellowish-green. Intranodes glabrous, or grayish-white-downy when young. Sheath scars swollen. Leaf blades linear-lanceolate to oblong-lanceolate, variable in size, usually 7 to 21 cm long, 1 to 2.8 cm wide; acuminate at apex, cuneate at base, oblique and asymmetrical; glabrous above, scabrous along midrib and on margins. Leaf sheath glabrous, striate. Leaf ligules short, pectinate, puberulent.

Distributed mainly in Guangdong and Guangxi Provinces. Tender.

Lingnania chungii var. *petille* Wen, var. nov.

Culms 3 to 4 m tall, 1 cm in diameter. Young culms green, glabrous, not pruinose. Culm wall thin and crisp. Culm sheath glaucous and sparse-hairy, but denser near base; concave at apex. Sheath auricles absent. Sheath ligules very short, sparsely long-ciliate at apex. Sheath blades about as broad as ⅓ of the sheath ligules. Sheath blades reflexed, lanceolate, downy on both surfaces. Leaves 8 to 10 on slender and glabrous twigs. Leaf sheaths 4 to 5 cm long, downy. Leaf auricles prominent, long-ciliate on rims. Leaf blades lanceolate, 9 to 16 cm long, 1.3 to 1.8 cm wide, downy on both surfaces.

Distributed in Fujian Province. Tolerates light frost.

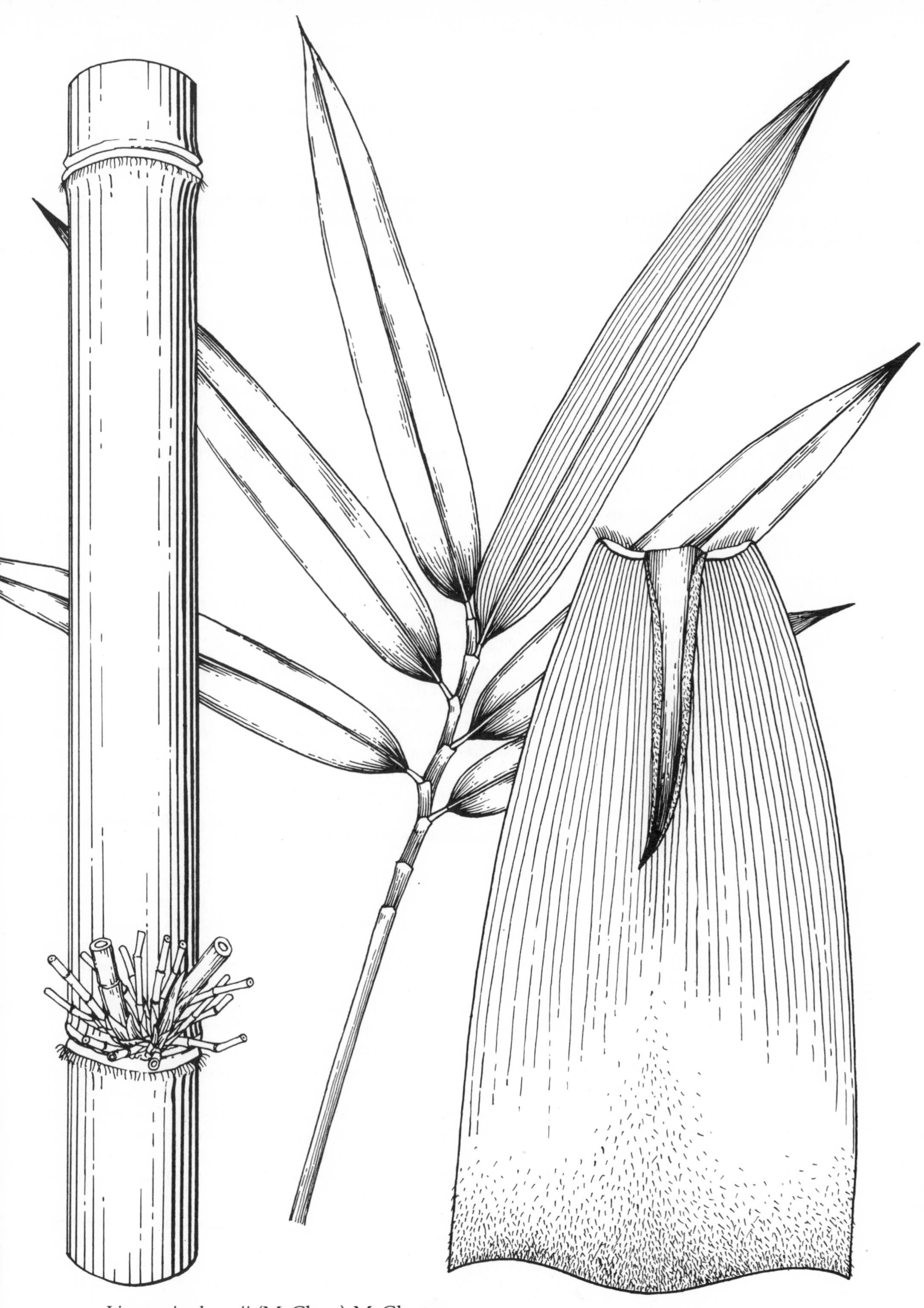

Lingnania chungii (McClure) McClure

XXIII. *Chimonocalamus* Hsueh et Yi, gen. nov.

Medium to small bamboos. Rhizomes sympodial. Culms cespitose. A girdle of aerial root thorns on nodes from midculm up. Internodes cylindrical, slightly compressed on branching side, with longitudinal ribs and grooves. Yellow essential oil inside lumen of culms. Nodal ridges slightly swollen or swollen. Culm sheaths deciduous, usually longer than internodes, papery to thin-leathery, brown-appressed-setulose on abaxial surface. Sheath auricles obsolescent. Sheath ligules prominent, fimbriate-ciliate. Sheath blades linear to lanceolate. Primary branches 3 or many on nodes. Branch nodes markedly swollen.

KEY to the Spp. of *Chimonocalamus* Native to China

1a. Leaf blades narrow, less than 1.3 cm wide. Lateral veins 2 to 4 pairs. Sheath blades over 3 cm long. 2

1b. Leaf blades broad, over 1.5 cm wide. Lateral veins 5 to 6 pairs. Sheath blades less than 3 cm long, twisting. .... (1) *Ch. tortuosus*

2a. Small to medium clumped bamboos. Culms over 3 m tall, 1 to 6 cm in diameter. Internodes slightly compressed on branching side, over 16 cm long. Culm sheath base over 4 cm wide. Primary branches 3 or many on nodes. 3

2b. Shrub-like bamboos. Rhizomes with somewhat elongated rhizome necks. Culms not closely clumped. Culms not over 3 m tall, 0.5 to 1.5 cm in diameter. Internodes cylindrical, usually not over 16 cm long. Culm sheath base 3 to 4 cm wide. Primary branches many on nodes..... (2) *Ch. dumosus*

3a. The top of culm sheath proper 2 to 4 cm wide, with ligule-like protruding apex. Sheath proper base 1 to 2 cm wide, glabrous. Aerial root thorns short and separated on nodes of mid- to lower-culm only. 4

3b. The top of culm sheath proper less than 1.5 cm wide, truncate, slightly convex or depressed at apex. Sheath blades slender, less than 1 cm wide at base, puberulent. Aerial root thorns usually closely clustered or separated, also found on lower nodes of primary branches (with the exception of *Ch. longiusculus*). 5

4a. Culms slightly square, purplish-brown when young, muricate. Culm sheath proper densely brown-pannose-setose on abaxial surface. The apex of sheath proper asymmetrical. Leaf blades grayish-green beneath. Oral setae on leaf sheath prominent. .... (3) *Ch. delicatus*

4b. Culms cylindrical, grayish-green when young as gray-farinose, smooth. Culm sheath proper not densely pannose-setose. The apex of sheath proper symmetrical. Leaf blades uniform in color of both surfaces. Leaf sheath oral setae none, or rarely 1 to 2 only. .... (4) *Ch. pallens*

5a. Culm sheaths brown-maculate, depressed at apex. The apex of sheath ligules fimbriate-lacinulate; lacinules 5 to 8 mm long. .... (5) *Ch. fimbriatus*

5b. Culm sheaths not maculate (sometimes lineate), not prominently depressed at apex. Lacinules on sheath ligules not over 5 mm long. 6

6a. Culm sheaths densely fugaciously verruciform-setose on abaxial surface and rims. Aerial root thorns short and blunt, closely set and sometimes united. .... (6) *Ch. montanus*

6b. Culm sheaths sparsely setose, not verruciform, truncate or slightly convex at apex, no setae on either side. Aerial root thorns longer, separated from each other. 7

7a. The apex of sheath proper 11 to 13 mm wide, sheath ligules 7 to 12 mm long. Sheath scars hairy. Aerial root thorns many, also on nodes of lower primary branches. Primary branches 3 on nodes. .... (7) *Ch. makuanensis*

7b. The apex of sheath proper 4 mm wide, sheath ligules less than 1.5 mm long. Sheath scars glabrous. Aerial root thorns sparse, not found on branch nodes. Primary branches many on nodes. .... (8) *Ch. longiusculus*

1. *Chimonocalamus tortuosus* Hsueh et Yi, sp. nov.

Rhizomes sympodial, culms cespitose. Culms 6 to 10 m tall, 1 to 3.5 (5) cm in diameter, with 35 to 54 internodes. Internodes 18 to 22 (28) cm on mid-culm nodes, cylindrical, yellowish-green; with gray-minute-chaetae on upper culm. Culm sheaths persistent, long-triangular, 2 times longer than internodes, densely brownish-black or yellowish-brown-setose except the basal part on abaxial surface, densely brownish-black or yellowish-brown-ciliate on rims. Sheath auricles tubercular, purplish-red, grayish-yellow-ciliate. Sheath blades small, 3 to 30 mm long, triangular or lanceolate, purple, upright. Sheath ligules dark purple, triangular, grayish-yellow ciliate. Sheath scars slightly swollen, densely grayish-yellow- or yellowish-brown-villous when young, glabrescent later. Aerial root thorns usually found on nodes below branching nodes. Primary branches 2 to 9 on nodes, spreading, 35 to 75 cm long. Leaves 3 to 7 on twigs. Leaf blades lanceolate, papery, 12 to 20 cm long, 1.1 to 2.4 cm wide, serrulate and scabrous on rims; green above, pale green beneath, glabrous. Transverse veinlets indistinct.

Distributed at an altitude of 1700 to 2200 m in Xizang (Tibet) Autonomous Region. Tender.

2. *Chimonocalamus dumosus* Hsueh et Yi, sp. nov.

Culms 1.5 to 3 m tall, 0.5 to 1.5 cm in diameter with 16 to 20 internodes. Young culms sparsely gray-farinose, glabrous. Internodes 16 cm long, cylindrical, without longituidnal rib on branching side; nearly solid only near culm base. Sheath scars glabrous. Nodal ridges prominently swollen. Culm sheaths papery, gray or yellowish-gray when young, appressed-setulose (yellowish-brown) on abaxial surface. Sheath proper arcuate or ligulate at apex. Sheath auricles absent. Sheath blades slender, upright. Sheath ligules 1.5 mm long, irregularly dentate-incised. Primary branches many, slender, green or purplish-brown. Leaves 3 to 7 on twigs. Leaf blades 3 to 16 cm long, 0.3 to 1.2 cm wide. Lateral veins 2 to 4 pairs.

Distributed at an altitude of 1500 m in Xichou Prefecture of Yunnan Province. Tender.

Chimonocalamus dimosus var. *pymgaeus* Hsueh et Yi, var. nov.

Small bamboo, 3 to 7 cm in diameter. Internodes 8 to 13 cm long. Aerial root thorns blunt, closely-set, and often united. Sheath callus remains. Culm sheath proper nearly truncate at apex. Leaf blades 5 to 11 cm long, 5 to 9 mm wide.

Distributed in Gengma Prefecture of Yunnan Province. Tender.

3. *Chimonocalamus delicatus* Hsueh et Yi, sp. nov.

Culms 8 to 10 m tall, 4 (8) cm in diameter; purplish-brown and muricate when young, light yellow and becoming smooth when old. Internodes cylindrical or slightly square, 20 to 22 (30) cm long. Nodal ridges swollen with a narrow rib. Sheath scars glabrous. Culm sheath coriaceous and crustaceous, long-rectangular, 20 to 45 cm long, 14 cm wide at base, narrowed to the 2 to 4 cm wide apex. Sheath auricles absent or obsolescent, with prominent oral setae. Sheath ligules pale-yellowish-brown, lucid, irregularly dentate-incised, 4 cm long. Sheath blades lorate-lanceolate, 5 to 17 cm long, upright. Branching high. Primary branches 3, fistulose or nearly solid, sometimes with a few aerial root thorns on nodes. Leaf blades long-lanceolate, 10 to 16 cm long, 6 to 13 mm wide, 5 to 10 mm long-awned at apex; green above, grayish-green beneath, glabrous. Lateral veins 3 to 4 pairs. Transverse veinlets distinct.

Distributed at an altitude of 1400–2000 m in Jinping Prefecture of Yunnan Province. Tender.

4. *Chimonocalamus pallens* Hsueh et Yi, sp. nov.

Culms 5 to 8 m tall, 2 to 5 cm in diameter, with 42 internodes. Internodes 12 to 29 cm long, slightly compressed on branching side, without distinct ribs and grooves. Nodal ridges swollen with a narrow rib. Sheath callus exserted when young, slightly hairy on callus and under nodes. Aerial root thorns on nodes below branching nodes, short and separated. Culm sheaths thin-leathery, indistinctly pale-brown-setulose, also pale-brown-maculate; densely puberulent at base; longitudinal veins distinct; ligulate at apex, 1.5 cm long. Sheath ligules dark brown, lucid, irregularly undulate-incised. Sheath blades lorate-lanceolate. Branching high, usually from the 10th node up. Primary branches 3 on nodes. Secondary branches also 3 on branch nodes. Leaves usually 6 on twigs. Leaf blades 13 cm long, 1.5 cm wide, 3 mm long-awned at apex, green on both surfaces, glabrous. Lateral veins 3 to 4 pairs. Transverse veinlets distinct.

Cultivated in Jinping Prefecture of Yunnan Province. Tender.

5. *Chimonocalamus fimbriatus* Hsueh et Yi, sp. nov.

Culms 5 to 8 m tall, 2 to 5 cm in diameter, with about 33 internodes. Young culms dark green or somewhat purplish, sparsely white-setulose and pubescent, fugaciously. Internodes cylindrical, slightly compressed at the base on branching side. Sheath scars and nodal ridges slightly swollen. Aerial root thorns, 7 to 14 mm long, up to more

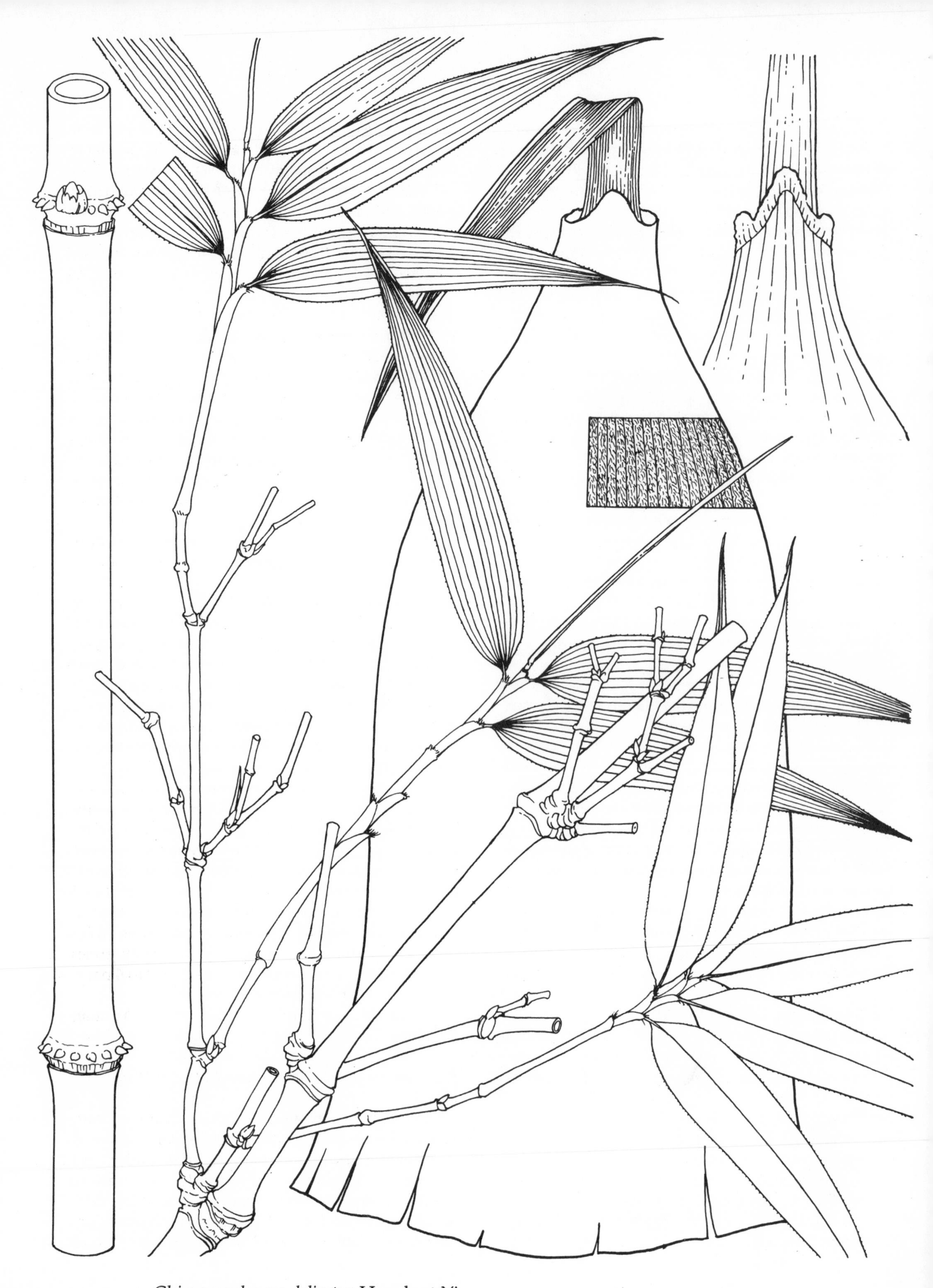

Chimonocalamus delicatus Hsueh et Yi

than 30, densely arranged and usually united at lower part. Culm sheaths thin-leathery, longer than internodes, abscising early, green with purplish-red shade; brown-maculate and brown-appressed-setose on abaxial surface; depressed at apex. Sheath ligules prominent, fimbricate-ciliate at apex; sheath blades upright or reflexed. Primary branches 3 or many on nodes. Leaves 3 to 5 on twigs. Leaf blades 5 to 15 cm long, 5 to 11 mm wide, 1 cm long-awned at apex. Lateral veins 3 to 4 pairs.

Distributed on southern Yunnan. Tender.

6. *Chimonocalamus montanus* Hsueh et Yi, sp. nov.

Culms 5 m tall, 1.5 cm in diameter. Internodes 33 cm long, cylindrical, slightly compressed at the base on branching side, with insignificant ribs and grooves, glabrous. Sheath scars slightly swollen, glabrous. Nodal ridges ribbed and swollen. Aerial root thorns grow up to several nodes of branches, 5 mm long, blunt at apex, closely set and usually united. Culm sheaths densely strewn with dark-brown verruciform and fugacious bristles on the abaxial surface, setose on rims; truncate or retuse at apex. Sheath blades slender-lanceolate, densely tomentose, involute on both sides, aristate at apex. Sheath ligules truncate at apex, 2 mm long, dark brown, ciliate at apex. Primary branches 3 on nodes. Leaves 2 to 4 on twigs. Leaf blades 14 cm long, 1 cm wide, with a 8 mm long-awned apex. Lateral veins 3 pairs.

Distributed at an altitude of 1740 m in southern Yunnan Province. Tender.

7. *Chimonocalamus makuanensis* Hsueh et Yi, sp. nov.

Culms 5 to 6 m tall, 1.5 to 2.5 cm in diameter, with 33 internodes. Internodes 10 to 27 cm long, pale green, brown-setulose when young, glabrous later, cylindrical, slightly compressed at base only on branching side. Sheath callus densely pale-brown-tomentose. Nodal ridges ribbed-swollen, puberulent. Aerial root thorns on every node of primary branches and also nodes on lower culm, 1 cm long, awl-shaped, closely set. Culm sheath longer than internodes, abscising early, yellow-striate when young, with appressed, needle-like brown bristles on abaxial surface. Sheath blades upright. Sheath ligules 7 to 12 mm long, fimbricate-incised at apex. Primary branches 3 on nodes, branch nodes swollen outside. Leaves 3 to 4 on twigs, 9 to 13 cm long, 9 to 13 mm wide, grayish-green beneath. Lateral veins 4 pairs.

Distributed at an altitude of 1700 m in Maguan Prefecture of Yunnan Province. Tender.

8. *Chimonocalamus longiusculus* Hsueh et Yi, sp. nov.

Culms 4 to 6 m tall, 1 to 2 cm in diameter, with 25 internodes. Culm wall thick, nearly solid at the internodes near culm base. Young culms dark green, densely white-minute-hairy under nodes. Internodes reaching 37 cm long, cylindrical, slightly compressed at the base of branching side, with a longitudinal groove, almost the length of the internode. Sheath callus glabrous. Nodal ridges prominently swollen. Aerial root thorns less than 10, thick and blunt, irregularly arranged on nodes. Culm sheaths abscising late, thick-papery, long-rectangular, appressed-brownish-yellow-setulose on abaxial surface. Sheath auricles small and asymmetrical, often obscure. Sheath blades linear, upright or spreading outward. Sheath ligules arcuate at apex, white-ciliolate. Primary branches 4 to 5 on every node, including 2 to 3 dominant branches, each with 3 to 5 twigs, pale purple in color. Leaves 3 to 5 on twigs. Leaf blades linear, 5 to 14 cm long, 5 to 9 mm wide, serrulate. Lateral veins 3 pairs. Transverse veinlets distinct.

Distributed at an altitude of 1650 m in Xichou Prefecture of Yunnan Province. Tender.

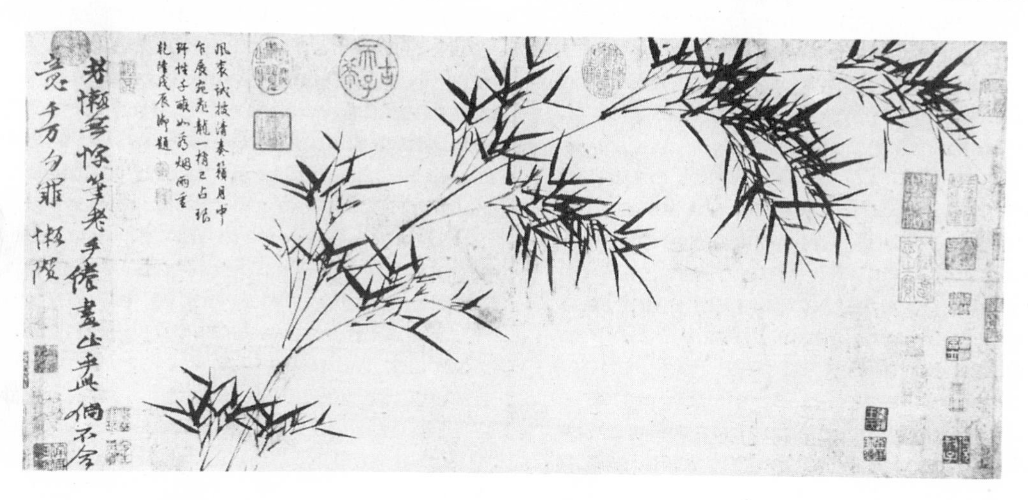

XXIV. *Dendrocalamus* Nees

Tall bamboos, rhizomes sympodial. Culms clumped, cylindrical, some spp. compressed or grooved on branching side; pruinose when young; erect, culm tops usually drooping. Primary branches many on nodes, not spiny. Sheath blades usually reflexed. Sheath auricles obsolescent.

KEY to the Spp. of *Dendrocalamus* Native to China

1a. Sheath auricles none. 2
1b. Sheath auricles distinct. Culm sheaths glabrous or somewhat brown-black-setulose on abaxial surfaces. Sheath blades glabrous. .... (1) *D. strictus*
2a. Upper culm sheath pruinose and sparsely blackish-brown-tufted-appressed-setose. Sheath ligules 6 mm long, undulate at apex. Sheath blades sparse-villous on abaxial surfaces, dark-brown-setulose between veins on adaxial surfaces. .... (2) *D. sinicus*
2b. Culm sheaths dark-yellowish-brown on abaxial surfaces, brownish-black-setose. Sheath ligules 2 to 4 mm long, arcuate. Sheath blades glabrous on abaxial surfaces; brownish-black-setose or glabrous on the middle of lower part of adaxial surfaces. .... (3) *D. tibeticus*

1. *Dendrocalamus strictus* (Roxb.) Nees

Culms 7 to 20 m tall, 2.5 to 7.5 cm in diameter. Culms hollow when grown in humid conditions, or nearly solid in dry conditions. Primary branches several on nodes, main branches larger than the others. Culm sheath glabrous; or somewhat brown- or black-setulose on abaxial surfaces. Sheath auricles distinct. Sheath blades triangular, glabrous. Leafy twigs slender, clustered. Leaves 3 to 13 on twigs. Leaf sheaths glabrous. Leaf ligules short, 0.5 mm long, truncate at apex. The size of leaf blades variable, 5 to 25 cm long, 1 to 3 cm wide, of thin texture, acuminate at apex, narrowed to a 1 mm long stalk. Leaf blades glabrous above, usually pubescent beneath, serrulate. Lateral veins 3 to 5 pairs. Transverse veinlets absent.

Distributed in Yunnan and Guangdong Provinces. Vertical distribution at about 1000 m. Tender.

2. *Dendrocalamus sinicus* Chia et J. L. Sun, sp. nov.

Giant bamboos. Culms 20 to 30 m tall, 20 to 30 cm or more in diameter. Culm wall 2 to 3 cm thick. Lower culms curving slightly, upper culms nearly erect, culm tops slightly drooping. Internodes cylindrical; but sometimes several internodes near culm base obliquely swollen, connected with each other in rhombus shape pattern. Normal lower culm internodes 17 to 22 cm long, thick-pruinose on the part covered by culm sheaths, also appressed-brown-setulose in longitudinal strips. A 3- to 4- mm zone yellowish-brown-sericeous on sheath scars when young. Nodes near culm base usually bearing short aerial roots. Primary branches several clustered on nodes up from ⅓ length of culms, all primary branches of even thickness, about 1 m long. Culm sheaths large, nearly persistent, of thick-leathery texture, usually longer than the related culm internodes. Leaf sheaths puberulent when young, glabrate later. Leaf auricles and oral setae obsolescent. Leaf ligules 1.5 to 2 mm long, sparse-denticulate. Leaf blades oblong-lanceolate to narrow-lanceolate, 20 to 40 cm long, 4 to 6.5 cm wide, very sparsely pilose or nearly glabrous. Lateral veins 10 to 13 pairs, transverse veinlets distinct.

Distributed at 500 to 1200 m in Gengma Prefecture of Yunnan Province. Tender.

3. *Dendrocalamus tibeticus* Hsueh et Yi, sp. nov.

Culms 15 to 25 m tall, 12 to 18 cm in diameter. Nodes 38 to 45 on culms. Culm tops drooping. Internodes 40 to 45 cm long; the longest up to 55 to 60 cm; while the shortest near base 28 to 30 cm. Culms green, glabrous, not pruinose or very sparsely pruinose. Grayish-white- or grayish-brown-downy zone both above and under nodes. Culm wall 0.6 to 1.2 cm thick, reaching 2.3 cm at basal nodes. Sheath scars swollen, glabrous. Nodal ridges not swollen. Intranodes grayish-white or grayish-brown-downy. The basal 3 to 7 nodes densely clustered with aerial roots. Primary branches many, clustered, usually branching from the 14th node upwards. Leaves 5 to 8 on twigs. Leaf auricles absent. Leaf ligules truncate at apex, glabrous; pruinose on abaxial surfaces. Leaf blades broad-lanceolate, 10 to 32 cm long, 2.2 to 4.5 cm wide; acuminate at apex; obtuse or broad cuneate at base; glabrous; green above, pale green beneath; serrulate on one side only, slightly scabrous, the other side nearly smooth. Lateral veins 8 to 13 pairs. Transverse veinlets distinct.

Distributed at an altitude of 1220 m in Xizang Autonomous Region (Tibet). Tender.

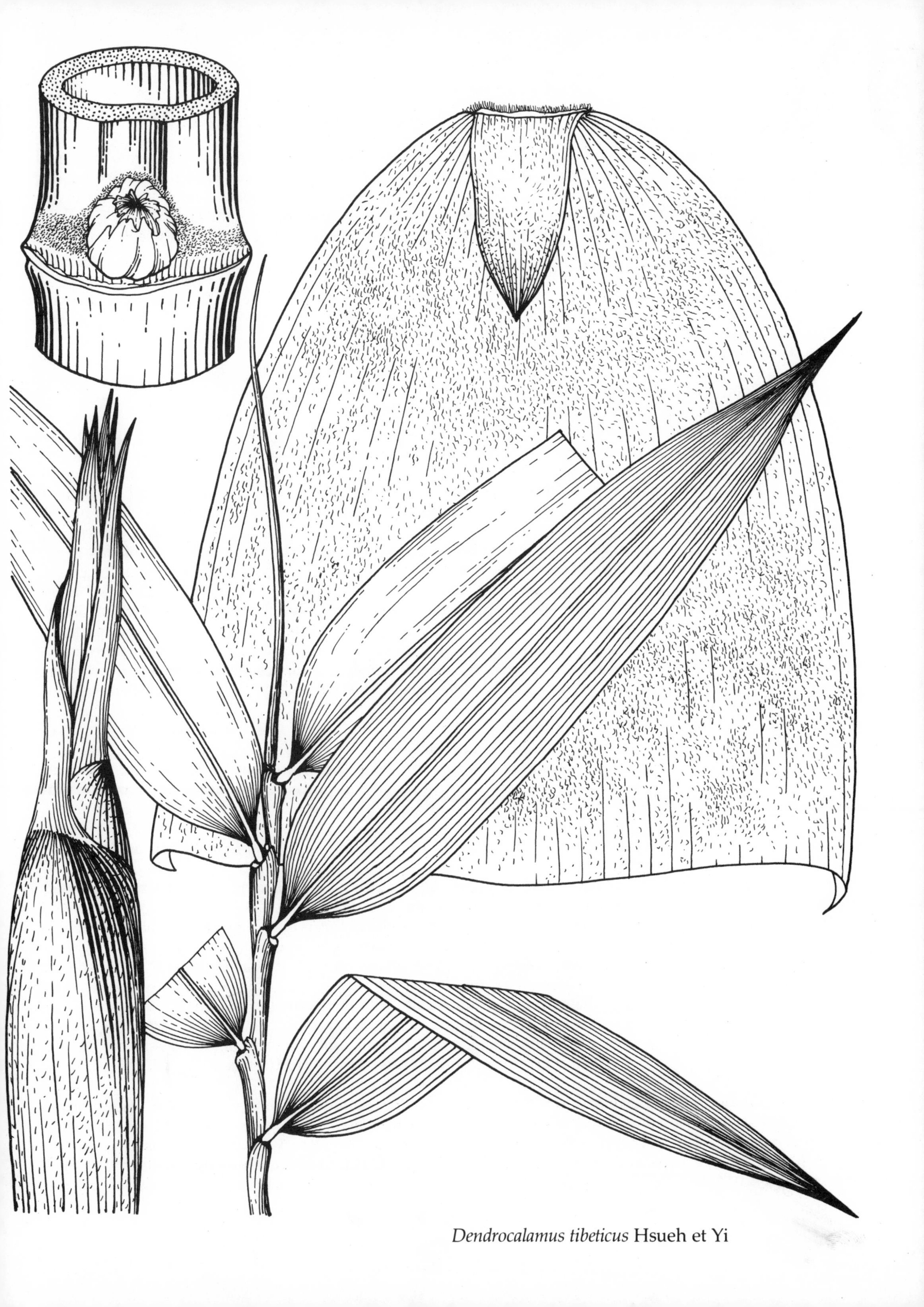

Dendrocalamus tibeticus Hsueh et Yi

XXV. *Sinocalamus* McClure

Tall bamboos; rhizomes sympodial; culms clumped, spineless; the top of culm arching or drooping. Internodes cylindrical. Culm sheaths deciduous, broad-obtuse at apex, often brown-setulose on abaxial surfaces. Sheath auricles obsolescent or absent; ligules prominent, sometimes long-ciliate. Sheath blades narrow and small, readily reflexed. Primary branches many on nodes, the central branch thicker and longer than the others. Leaf blades usually large, broad-lanceolate to oblong-lanceolate, sometimes with transverse veinlets.

KEY to the Spp. of *Sinocalamus* Native to China

1a. Culms about 10 m tall, 2 to 7 cm in diameter. 2
1b. Culms tall (except *S. minor*), usually over 15 m, more than 10 cm in diameter. 5
  2a. Culm tops slant-spreading, not arching. 3
  2b. Culm tops arching to drooping. 4
    3a. Culm sheaths glabrous and lucid on abaxial surface. Sheath blades usually upright. Sheath ligules 1 mm long, entire. Sheath auricles small, prominently ciliate. Leaf auricles prominent. .... (1) *S. oldhami*
    3b. Culm sheaths appressed-black-setose on abaxial surface. Sheath blades small, upright or reflexed. Sheath ligules 3 to 4 mm long, serrulate or denticulate. Sheath auricles and leaf auricles none. .... (2) *S. bicicatricatus*
      4a. Culm sheaths densely brownish-black-setose on abaxial surface. Sheath blades ovate-lanceolate, upright or reflexed. Sheath ligules prominent, over 10 mm long, including cilia on apex. Sheath auricles absent. .... (3) *S. affinis*
      4b. Culm sheaths glabrous or appressed-brown-setose only at base on abaxial surface. Sheath blades triangular-lanceolate or triangular-ovate, upright. Sheath ligules nearly entire or serrulate at apex. Sheath auricles remain. .... (4) *S. vario-striatus*
        5a. Culm sheath hard and brittle, smooth on abaxial surface, dark-brown-bristly at earlier stage, then glabrate. Leaf blades large, 15 to 50 cm long, 4 to 13 cm wide. Lateral veins 11 to 16 pairs. 6
        5b. Culm sheaths of leathery texture, with longitudinal lines on abaxial surfaces, usually setulose. Leaf blades 11 to 34 cm long, 1.5 to 5.5 cm wide. Lateral veins 5 to 13 pairs. ... 8
          6a. Leaf ligules truncate at apex, 1 mm long. Transverse veinlets beneath leaf blades not prominent. The base of sheath blades decurrent into auricles, glabrous, brown, waxy-powdery; sheath ligules hard, black, serrulate. Leaf sheaths glabrous. .... (5) *S. giganteus*
          6b. Leaf ligules convex at apex, 1.5 to 2 mm long. Transverse veinlets beneath leaf blades prominent. 7
            7a. Culms 20 to 25 m tall. Upper part of leaf sheaths covered with appressed, yellowish-brown hair. Lateral veins 11 to 15 pairs. .... (6) *S. latiflorus*
            7b. Culms 1.5 to 8 m tall. Leaf sheaths glabrous or slightly hairy on abaxial surfaces and margins. Lateral veins 12 pairs. .... (7) *S. minor*
              8a. Leaf ligules short. The lower part of abaxial surface of culm sheaths appressed-bristly; while the upper part sparsely pale-purplish-bristly. Lateral veins 10 to 13 pairs, with secondary lateral veins 8 to 9 between each pair of lateral veins. .... (8) *S. edulis*
              8b. Leaf ligules longer, up to 7 mm. Only the base of abaxial surfaces of culm sheath appressed-bristly. Lateral veins 5 to 10 pairs. ...... (9) *S. beecheyanus*

1. *Sinocalamus oldhami* (Munro) McClure

Culms 6 to 9 (sometimes up to 20) m tall, 5 to 8 cm in diameter. Primary branches many on nodes, including 3 large ones and some smaller ones. Leaves 7 to 15 on twigs. Leaf sheaths 7 to 15 cm long, setulose, glabrate later. Leaf blades broad-lanceolate, 12 to 30 cm long, 2.5 to 6.2 cm wide; lateral veins 9 to 14 pairs, with transverse veinlets.

Distributed in southern Zhejiang Province, Fujian, Guangdong (including Hainan Island),

Guangxi, and Taiwan Provinces. The vertical distribution up to 700 m. Tender.

2. ***Sinocalamus bicicatricatus*** W. T. Lin, sp. nov.

Culms nearly erect, over 10 m tall, 5.5 to 6.3 cm in diameter. Nodes slightly swollen or those on lower culm prominently swollen. Sheath scars narrow and slightly swollen, those on lower culms consist of two girdles of horizontal lines. Internodes 20 to 36 cm long, cylindrical, straight and glabrate, green. Culm sheaths deciduous, falcate or rounded at apex. Primary branches usually 3, sometimes 5 on nodes, the main branch thicker and longer than the others. Leaves 6 to 9 on twigs. Leaf blades long-elliptic-lanceolate, 9 to 22 cm long, 2 to 4 cm wide, acuminate at apex, rounded or cuneate at base. Lateral veins 6 to 11 pairs. Transverse veinlets distinct on the lower surface.

Distributed in Hainan Island of Guangdong Province. Tender.

3. ***Sinocalamus affinis*** (Rendle) McClure

Culms 5 to 10 m tall, 3 to 6 cm in diameter. Internodes appressed-bristly, grayish-white or grayish-brown, glabrate later, but small pores or minute warts remain after shedding bristles. Intranodes of lower culm usually white-hairy. Primary branches over 20 on midculm nodes, in semi-whorl pattern. Leaves several to over 10 on twigs. Leaf blades thin, 10 to 30 cm long, 1 to 3 cm wide; lateral veins 5 to 10 pairs.

Distributed in Sichuan, Guizhou, Guangxi, Guangdong, Hunan, western Hubei, southwestern Shaanxi, and southern Gansu Provinces. Tolerates a minimum temperature of −15°C.

4. ***Sinocalamus vario-striatus*** W. T. Lin, sp. nov.

Culms nearly erect, 5 to 12 m tall, 4 to 7 cm in diameter. Internodes cylindrical or somewhat swollen, with appressed pubescence in longitudinal lines or pale-purple-striate when young, but glabrous and yellowish-white striate when mature. Nodes slightly swollen. Intranodes on lower culm with a girdle of appressed white pubescence. Culm sheaths deciduous. Sheath auricles nearly equal, long elliptic. Oral setae 4 mm long. Sheath ligules 3 to 9 mm long. Sheath blades upright, triangular-lanceolate. Primary branches many on nodes, one main branch dominating with two less dominant, while the others short and small clustered together. Leaves 8 to 12 on twigs. Leaf blades linear-lanceolate, 13 to 26 cm long, 1.6 to 3 cm wide. Lateral veins 6 to 10 pairs.

Distributed in Guangzhou of Guangdong Province. Tender.

5. ***Sinocalamus giganteus*** (Wall.) Keng f.

Culms 25 to 30 m tall, 15 to 25 cm in diameter. Internodes dark green or grayish-green, glaucous when young, 40 cm long. Bristly under sheath scars. Culm sheaths large, abscising early, hard. Branching only on upper culm nodes. Leaves about 8 on twigs. Leaf blades on strong twigs often 30 to 55 cm long, 10 to 11 cm wide; but those on weak growth 15 cm long, 2.5 to 4 cm wide; oblong, asymmetrical longitudinally; acuminate with a slightly twisted sharp apex; puberulent beneath when young. Lateral veins 9 to 12 pairs, transverse veinlets not distinct.

Distributed in Yunnan Province. Tender.

6. ***Sinocalamus latiflorus*** (Munro) McClure

Culms 15 to 20 (25) m tall, 10 to 20 (30) cm in diameter. Internodes 30 to 45 cm long. Leaves 7 to 10 on twigs. Leaf blades long-lanceolate to broad-lanceolate or elliptic-lanceolate, 15 to 40 (50) cm long, 4 to 8 (14) cm wide, glabrous above, midrib prominent on the back, serrulate, transverse veinlets prominent.

Distributed in southern and southwestern China, Viet Nam, and Burma. Tender.

Sinocalamus latiflorus var. *magnus* Wen, var. nov.

Culms erect, 18 m tall, 16 cm in diameter. Internodes 36 cm long. Aerial roots on lower culm nodes. Branching from 5 to 9 m up. Branches puberulent when young, glabrate later. Leaves 6 to 9 on twigs. Lateral veins 12 pairs. Transverse veinlets distinct.

Distributed in Fujian Province. Tolerates a minimum temperature of −3°C.

(The species branching low, from 1 m up; culms arching; young branches glabrous.)

7. ***Sinocalamus minor*** McClure

Culms 1.5 to 8 m tall. Internodes cylindrical, glabrous, glossy. Branch nodal ridges swollen. Single twig on branch nodes. Leaf sheaths lineate. Leaf auricles none; leaf ligules 1.5 mm long, glabrous or nearly glabrous on abaxial surface. Leaf blades oblong-lanceolate, 18 cm long, 55 mm wide, glabrous; paler beneath; transverse veinlets distinct on the lower surface.

Distributed in plains and along watersides of Guangdong and Guangxi Provinces. Tender.

Sinocalamus minor var. *amoenus* Q. H. Dai et C. F. Huang

Culms smaller. Internodes pale yellow with green strips.

Distributed in southern Guangxi Provinces. Tolerates lime. Ornamental. Tender.

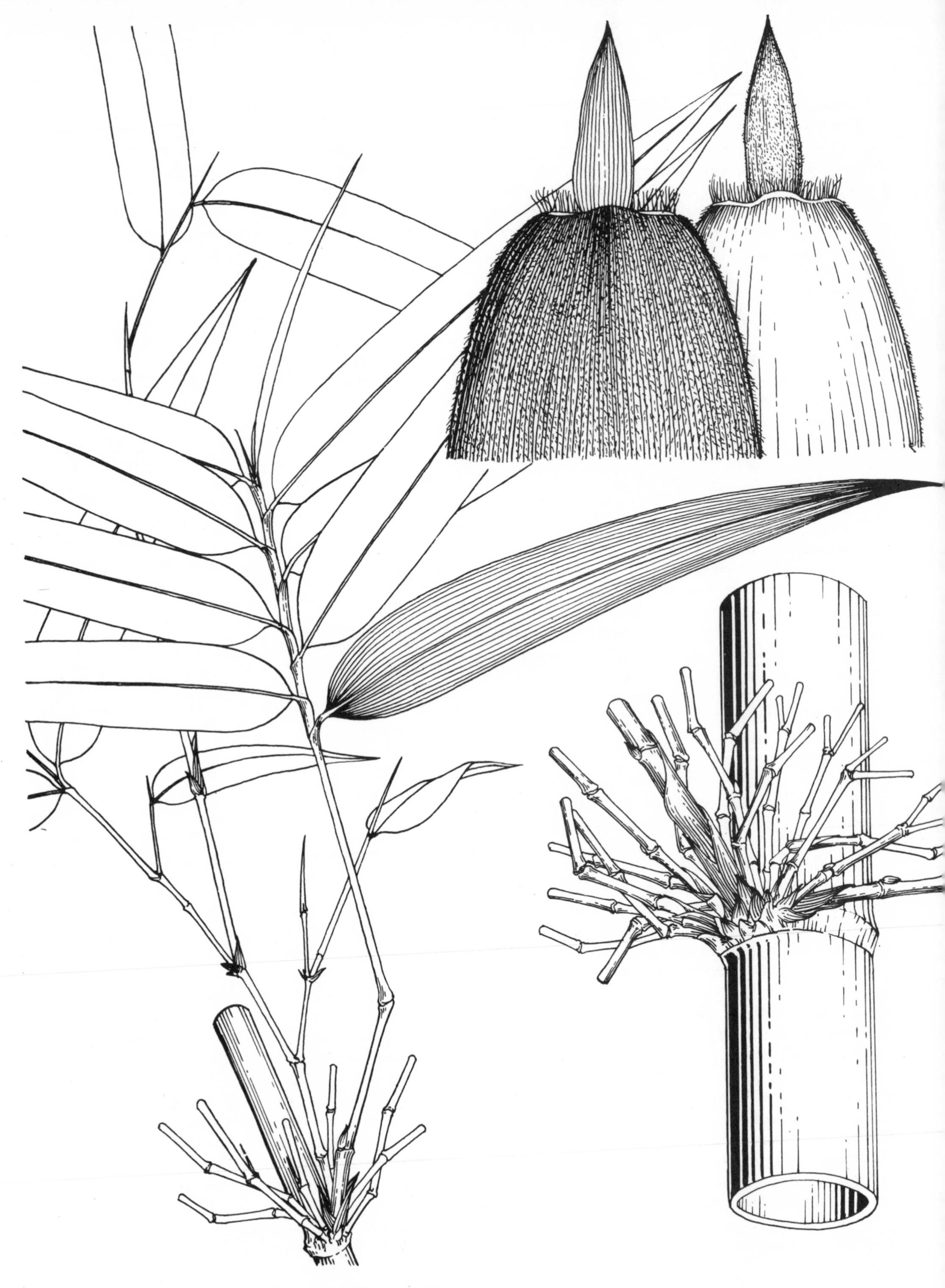

Sinocalamus affinis (Rendle) McClure

8. *Sinocalamus edulis* (Odashima) Keng f.

Culms 20 m tall, 7.5 to 13 cm in diameter. Internodes 20 to 35 cm long, pale green. Primary branches many on nodes, growing into a bundle. Leaves 10 to 12 on twigs. Leafy twigs usually slender and weak. Leaf sheaths glabrous, lineate. Leaf auricles small. Leaf ligules prominent, truncate at apex. Leaf blades oblong-lanceolate to long-narrow-lanceolate; 20 to 34 cm long, 3 to 5 cm wide; glabrous above; grayish and bristly beneath.

Distributed in Taiwan Province. Tender.

9. *Sinocalamus beecheyanus* (Munro) McClure

Culms 8 to 12 m tall, 5.5 to 9 cm in diameter. Culm tops arching. Internodes 27 to 31 cm long, young culms glaucous, especially from midculm upwards. Branching high. Primary branches clustered on nodes. Leaves 6 to 14 on twigs, of unequal sizes. Leaf blades oblong-lanceolate, 5 to 29 cm long, 1.1 to 5.5 cm wide, acuminate with an awl-shaped apex; glabrous above, somewhat scabrid beneath; lateral veins 5 to 10 pairs, transverse veinlets distinct.

Distributed in plains or hilly areas of southern China. Commonly cultivated. Tender.

Sinocalamus beecheyanus var. *pubescens* P. F. Li

Young culms hairy and hairy zones above nodes from midculm downwards. Internodes usually shorter than the sp. Culm sheaths usually with diffuse, blackish-brown, appressed and acropetal setae on abaxial surface. The base of sheath blade narrower than, and the sheath ligule longer than the sp. Leaf sheath usually hairy, leaf ligules longer than the sp.

Distributed in Guangdong and Guangxi Provinces. Tender.

XXVI. *Dinochloa* Buse

Giant climbing bamboos. Rhizomes sympodial. Culms cespitose, cylindrical, zigzag. Nodes swollen. Primary branches many clustered on nodes, the central one thicker and longer than the others, and the others of equal size. Culm sheaths deciduous. Sheath blades long, reflexed. Sheath auricles not prominent. Leaf blades usually large, somewhat soft. Transverse veinlets between lateral veins not distinct.

KEY to the Spp. of *Dinochloa* Native to China

1a. Sheath auricles remain. 2
1b. Sheath auricles absent. .... (1) *D. orenuda*
  2a. Sheath auricles reflexed, brown, setae 1.5 to 2.5 cm long on rims. Internodes densely yellowish-white-sericeous. Culm sheaths appressed-brown-pubescent on abaxial surface. ...... (2) *D. puberula*
  2b. Sheath auricles not reflexed, but erect along the apex of sheath proper, usually not hairy on rims. Culms glabrous except yellowish-white-sericeous on intranodes only. .... (3) *D. utilis*

1. *Dinochloa orenuda* McClure

Culms 9 to 20 m tall, 2.2 to 3.5 cm in diameter. Internodes slightly curved at nodes, densely appressed-yellowish-white-sericeous on intranodes and under nodes. Leaf blades lanceolate to oblong-lanceolate, 7 to 18 cm long, 1.2 to 3 cm wide; acuminate and subulate at apex, rounded or broad-cuneate at base; glabrous or pubescent beneath; scabrid on rim. Leaf ligules arcuate at apex, serrulate.

Distributed in Hainan Island of Guangdong Province. Tender.

2. *Dinochloa puberula* McClure

Culms 10 to 30 m long, 2 to 3.5 cm in diameter. Internodes slightly curving, lanose-pubescent when young. Sheath scars prominently swollen. Culm sheaths densely brown-pubescent when young, glabrous later. Primary branches many, clustered, short and small, except the central one prominently stronger and longer. Leaves 13 on twigs. Leaf blades oblong-lanceolate to lanceolate, 5 to 25 cm long, 1.2 to 4 cm wide; acuminate and apiculate at apex, glabrous or sometimes pilosulose beneath.

Distributed in Hainan Island of Guangdong Province. Tender.

3. *Dinochloa utilis* McClure

Culms 9 m tall, 4 to 5 cm in diameter. Internodes strigose when young, denser on upper part. Leaf blades lanceolate, 7 to 13 cm long, 1 to 2 cm wide, acuminate and subulate at apex, rounded at base, glabrous, scabrous on rim. Leaf auricles slender, protruding, with several setae. Leaf ligules slightly arcuate at apex, entire.

Distributed in Hainan Island of Guangdong Province. Tender.

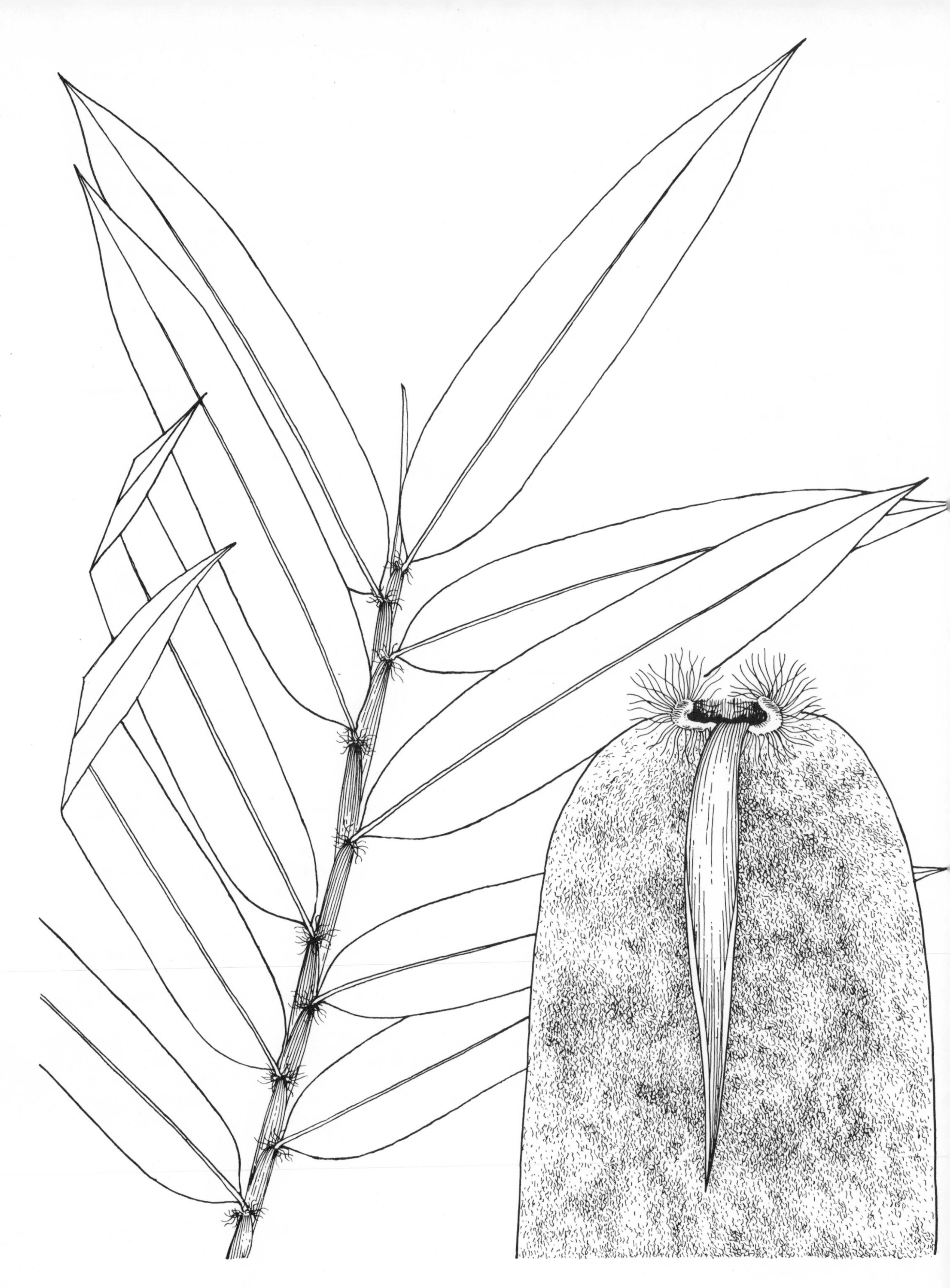

Dinochloa puberula McClure

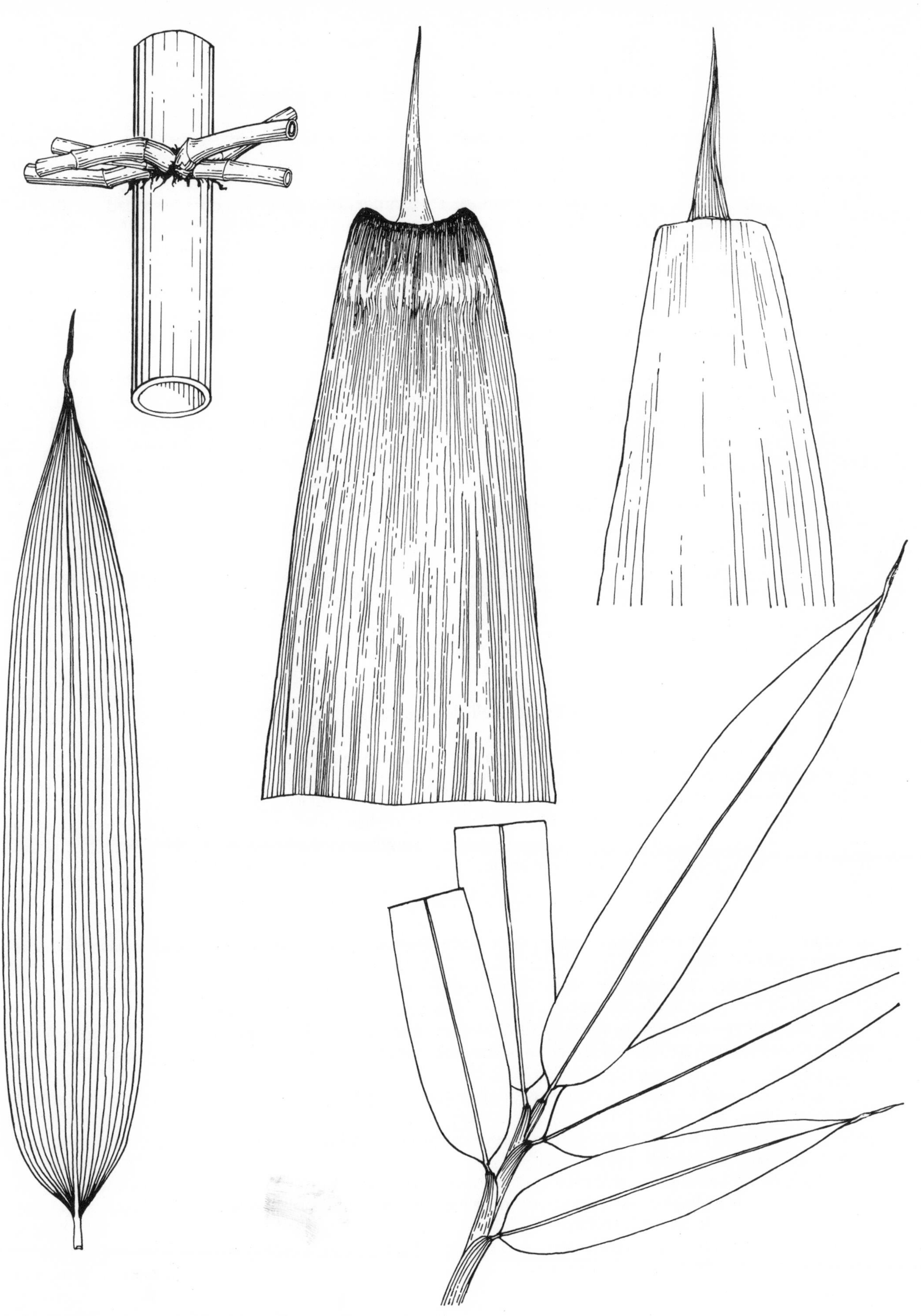

Leptocanna chinensis (Rendle) Chia et H.L. Fung

XXVII. *Leptocanna* Chia et H. L. Fung, gen. nov.

Shrub-like bamboo. Rhizomes sympodial. Culms cespitose, drooping or climbing at top. Internodes cylindrical, siliceous. Culm sheaths deciduous, but sheath callus remains; pruinose zone under nodes. Primary branches many, clustered on nodes, of even thickness. Culm sheaths thick-papery, crisp, siliceous on abaxial surface. Part near top of sheath proper transversely protruded on abaxial surface to a roundish-arcuate protuberance. Margins of sheath proper base decurrent to half-circular appendages. Sheath blades broad-linear, erect, corners on base porrect to extremely narrow-linear auricles. Oral setae obsolescent. Sheath ligules extremely narrow. Leaf sheaths striate. Leaf auricles and oral setae obscure. Leaf blades large. Transverse veinlets distinct.

Leptocanna chinensis (Rendle) Chia et H. L. Fung, comb. nov.

[*Schizostachyum chinense* (Rendle)]

Culms 5 to 8 m tall, 2 to 3 cm in diameter. Culm wall 2 to 3 mm thick. Internodes straight, 30 to 45 cm long; sparsely white-pubescent on upper part, glabrous when old; siliceous and scabrous. Branching from the 3rd node above culm base. Primary branches horizontally spreading, 80 to 100 cm long. Culm sheaths as long as ½ of internodes, purplish-red when young, stramineous when old. Culm sheaths nearly trapezoid; truncate or concave at apex; white-setulose when young on abaxial surface, fugacious, siliceous and scabrid when old. Sheath blades broad-linear; long-acuminate at apex, rims longitudinally incurved; the base of blade as broad as ⅓ of the top of sheath proper, porrect to extremely narrowed auricles. Sheath ligules 1 mm long, nearly entire. Leaf blades glabrous, lanceolate to oblong-lanceolate, 15 to 26 cm long, 3 to 4.5 cm wide; long-acuminate and twisted-scabrous-apiculate at apex, roundish or broad-cunate at base; scabrous beneath. Lateral veins 7 to 9 pairs. Leaf stalks somewhat purplish-red, 5 mm long.

Distributed in Yunnan Province at an altitude of 1500 to 2500 m. Tender.

XXVIII. *Schizostachyum* Nees

Tall or shrub-like bamboos, sometimes climbing; rhizomes sympodial, culms clumped. Internodes cylindrical, usually slender, somewhat siliceous, sparse-hairy. Nodes not swollen. Culm sheaths abscising late, siliceous and striate on abaxial surfaces, bristly; oral setae prominent. Sheath blades reflexed, lorate-lanceolate. Sheath auricles obsolescent. Sheath ligules truncate at apex, serrulate, or fimbriate-ciliate. Primary branches slender, clustered on all culm nodes; the length and thickness of primary branches nearly even. Leaves broader.

KEY to the Spp. of *Schizostachyum* Native to China

1a. Culms erect, not climbing. Culm sheaths truncate at apex. At both corners of the sheath base no decurrent ligule-like appendages. 2

1b. Culms slanting in position, climbing. Uplifting rounded shoulders on both corners of the apex of culm sheaths. Decurrent ligule-like appendages at both corners of the sheath base. 3

2a. Rims of sheath ligules uniformly fibriate. Oral setae 12 mm long, smooth. . . . . (1) *S. pseudolima*

2b. Rims of sheath ligules slightly lobed or irregular-fimbriate. Oral setae 5 mm long, scabrous at base of hairs. . . . . (2) *S. funghomii*

3a. The rounded shoulders of sheath apex apparently higher than the conjunctive line of sheath blade and sheath proper. 4

3b. The rounded shoulders of sheath apex not apparently higher than the conjunctive line of sheath blade and sheath proper. . . . . (3) *S. dumetorum*

4a. Sheath ligules scabrous and uneven near rims, nearly glabrous. Leaf blades with sparse-short-bristle-like hairs above, nearly glabrous beneath, obtuse-serrulate, transverse veinlets visible on both surfaces. . . . . (4) *S. xinwuense*

4b. Sheath ligules fimbriate on rims. Leaf blades nearly entire, scabrous on both surfaces. Transverse veinlets absent. . . . . (5) *S. hainanense*

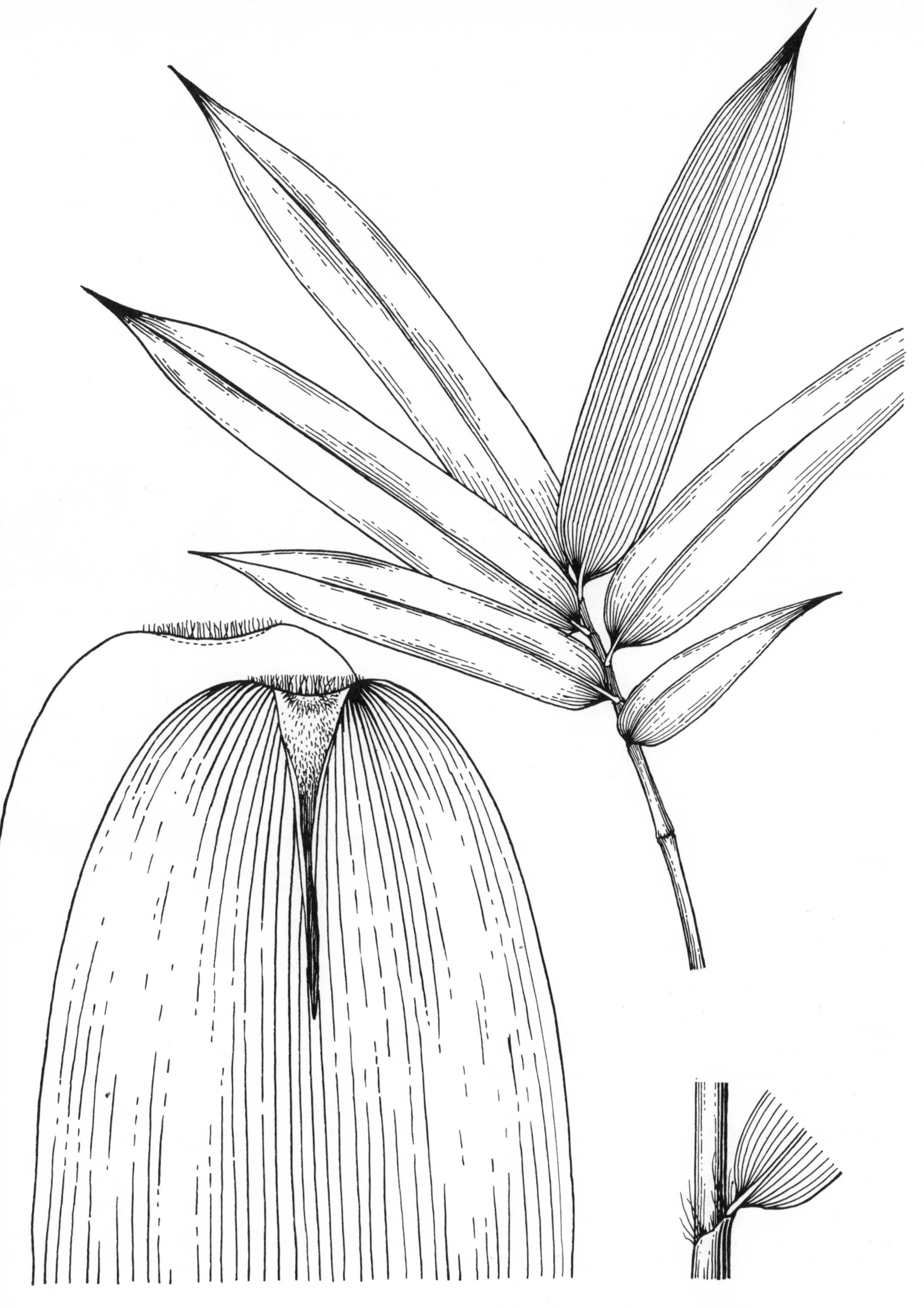

Schizostachyum funghomii McClure

1. *Schizostachyum pseudolima* McClure

Culms 10 m tall, about 4 cm in diameter, erect or nearly erect, culm tops drooping or climbing. Internodes 60 cm long or more, slightly curving. Culm wall thin, siliceous and tuberculate. Sheath scars swollen. Primary branches 50 cm long. Leaf blades oblong-lanceolate, 9 to 25 (33) cm long, 15 to 33 (48) mm wide; the upper surface scabrous along midrib and near margins, sometimes sparse-white-bristly; the lower surface pubescent, glabrate later; serrulate; lateral veins not prominent.

Distributed in Hainan Island of Guangdong Province. Tender.

2. *Schizostachyum funghomii* McClure

Culms 15 m tall, 4 to 6 cm in diameter. Internodes 40 cm long, scabrous and siliceous, appressed setulose at earlier stage, then glabrate and tuberculate. Leaf blades long-ovate to lanceolate, 10 to 25 cm long, 20 to 32 mm wide, glabrous and scabrous near margins above; puberulent beneath; serrulate; lateral veins 6 to 8 pairs.

Distributed in Guangdong and Guangxi Provinces. Cultivated for ornamental purposes. Tender.

3. *Schizostachyum dumetorum* (Hance) Munro

Culms 4 to 5 m tall, (sometimes up to 10 m), 1 to 1.5 cm in diameter. Internodes 30 cm long. Only those primary branches on nodes near culm base branching again. Sheath ligules glabrous, entire or undulate. Leaves 6 to 9 on twigs. Leaf blades scabrous above, nearly glabrous beneath, 5 to 13 (21) cm long, 8 to 15 (20) mm wide, serrulate, lateral veins 4 to 6 pairs. Transverse veinlets not prominent.

Distributed in Guangdong Province. Ornamental. Tender.

4. *Schizostachyum xinwuense* Wen et J. Y. Chin, sp. nov.

Culms 8 m tall, 1 cm in diameter. Internodes usually about 45 cm long, but the longest ones reaching 100 cm sometimes; white-appressed-sericeous, denser under nodes; not pruinose. Old culms glabrate. Sheath scars corky, smooth and glabrous. The length of culm sheaths about ½ of the length of internodes. Leaves 3 to 7 on twigs. Longitudinal veins on leaf sheaths not prominent. Leaf sheath glabrous. Leaf auricles absent. Oral setae erect. Leaf ligules truncate or arcuate at apex, scabrous. Leaf blades lanceolate to narrow lanceolate, 12 to 19 cm long, 7 to 20 mm wide, acuminate or acute at apex. Lateral veins 4 to 5 pairs.

Distributed in Jiangxi Province. Tender.

5. *Schizostachyum hainanense* Merrill ex McClure

Culms 5 to 17 m tall, 5 cm in diameter. Culm tops drooping or climbing. Internodes 75 cm long or more. Culm wall thin. Branches slender, less than 40 cm long. Leaf blades lanceolate to linear-lanceolate, 9 to 22 cm long, 1.2 to 2.5 cm wide. Leaf sheaths striate, white-appressed-setose.

Distributed in Hainan Island of Guangdong Province. Tender.

故宮角樓
癸亥
寫

4 THE GARDEN USE OF BAMBOO

The introduction of bamboos into Chinese gardens occurred at the very beginning of our garden history. It is native to the regions where the first emperors made their gardens. There were no reasons why bamboo should be left out and many for including it. It may prove interesting to our readers to retell a few stories about the importance of bamboos in Chinese gardens.

The First Emperor of the Qin Dynasty (259–210 B.C.) made one of China's most famous gardens, Yun Ming Tai (Terrace of Bright Cloud). Plants were collected from far and near to be grown there. It was the first example of a botanic garden designed to display a wide variety of plants. Among them was the 'White Bamboo' from Yun Kan.

In the year 138 B.C., the Emperor of the Han Dynasty, Han Wu Ti, ordered the conversion of a huge area of about 100 square miles into a complex of hunting fields, forests, recreation grounds, palaces, and gardens. It was named Shan Lin Yuan (Royal Garden of Superb Wood). Seventy groups of buildings were located on the grounds, one of which was a complex of heated conservatories for growing subtropical plants. More than 2000 kinds of plants, including bamboos, were cultivated in the gardens and conservatories.

About 500 years later, Buddhism became the national religion of North China. A famous temple, Yun Lin Sheh (Temple of Cloud and Wood), was built in Loyang with more than 1000 rooms. It was surrounded by gardens in which were grown large plantings of pines, cypresses, bamboos, and sweet-smelling herbs. Numerous other temple gardens followed. The gardens of the temples were open to the common people and were enjoyed by a large number of them. The people came to regard 'Temple Meetings' like a festival. 'Temple Meetings' are usually scheduled on the birthdays of a well-known Buddhist deities. People come from far and near to pay homage and enjoy the setting. Hundreds of temporary stalls selling every kind of goods crowded the nearby streets. In this way millions of ordinary people came to appreciate and enjoy the beauty and tranquillity of gardens.

Other landscaped areas and gardens were also open to the public in ancient China, i.e. places noted for inspiring natural scenery on public land or hills with some garden structures and roads provided by local officials, squires, Buddhist or Taoist temples. Bamboo was one of the popular ornamental plants used in these parks.

In the Tang Dynasty (618–905), many beautiful gardens were built in Sian (the Capital) and Lo Yang (the East Capital) by high-ranking officials, for their pleasure. Bamboo was given a prominent place in these private residential gardens. 'Place of excellent waters and bamboos' became the standard maxim of high praise for a garden in these times. The great poet, Tu Po (712–770.) after visiting one such garden wrote 10 poems in series. His first poem begins:

A famous garden lies near the green water,
Numerous bamboos shoot up to the blue sky.

He chose bamboo as representative of all ornamental plants.

No celebrated author of Chinese literature or poetry has written anything unfavorable about the bamboo. As a consequence and due to their immense influence, the Chinese people came to accept the idea that bamboo is a necessity in every garden. Even in the northern reaches of the Yellow River basin, where bamboos are difficult to maintain owing to severe winter weather, people continue to husband a few bamboo plants in the same way that they cherish their most precious possessions. It is told that in the Tang Dynasty, only one small cluster of bamboos was grown in the whole of

Beijing. They were only a few feet high and carefully protected from the weather. The gardener responsible for them was required to report daily on them to the owner.

Clumped bamboos are not really sufficiently hardy to survive in Sian, but some of them manage if carefully tended. The romantic emperor of the Tang Dynasty, Tang Ming Huang (685–762), had them grown near the Tai-Yi Lake in his palace. He used them to give his brothers, and by inference his subjects, a lesson illustrated by the growth habit of bamboo. He said that all the members of a family should stick together like the clumped bamboos did. This historic legend has been repeated down through the centuries to encourage familial devotion, and the bamboo as its symbol has been accorded the same respect.

Although difficult to grow and hence of limited use in the North, no garden in Central or Southern China would look complete without bamboos. The high ornamental value of bamboo deserves such respect and attention. Bamboo possesses great individual elegance. The stoloniferous species of bamboo build up a high green wall with a fine tex-

Bamboos soften the rigid lines of modern buildings and form a screen for a group of small pavilions at the center of Guilin.

ture, while the numerous twigs with their dense foliage drooping almost to the ground make the bamboo grove as thick and luxuriant as any shrubbery. The clumped bamboos stand by themselves in perfect grace. Their canes cling together supporting their gently drooping and spreading top. The whole clump looks like a huge basket. The skyline of a bamboo planting of any kind is in constant motion and quite different from that of trees.

The intellectuals of ancient China enjoyed bamboos, not only for their ornamental value, but also for the appealling sound of their twigs and foliage when the breeze rustled their tops. In Chinese it is described as the "Sound from Heaven", meaning the sound of Great Nature. As such it is indispensable to the proper order of all things.

An old book, *Nan Bu Yan Hua Lu* (Pleasure In The South), said to be written by Yang Shi-Gu in the early 7th century and made up of stories about the life of Sui Yan Ti (569–618, a tyrant of Sui Dynasty), casts some light on the role of The Sound of Great Nature. The Queen of Sui Yan Ti loved to listen to the rustling sound of bamboos near the pond which was close to her bedroom. When the bamboos died, she could not sleep. The emperor ordered that thin slices of jade formed as dragons be hung from the eaves overhanging her window until bamboo could be regrown. When the breeze came, the jade slices produced a sound just like the bamboos did, and the queen could once again sleep. This story could be an exaggeration, but it is quite possible that perfect stillness makes people uneasy. When roused from slumber in the depth of the night, the sound of the breeze passing through bamboo might well lead to a sense of peacefulness and put the sleeper at ease.

Through the centuries numerous Chinese artists have loved the bamboo and painted it. The central role of the bamboo theme in Chinese ink painting can be traced back about eleven hundred years. The classic history of Chinese art, *Tu Hui Bao Jian* (Record of Painters and Paintings), published in 1365 by Xia Wen-Yan, tells us how this theme began. In the Tang Dynasty, there was a Lady Li, from an eminent family of Western Sichuan. She was an accomplished writer, painter, and calligrapher. When General Kuo Chong-Tao overthrew the rebellious governer of Sichuan, she was forced to become the general's concubine. She was unhappy, as she found her husband a vulgar man. One night as she sat alone in her sitting room reflecting on her unhappy fate she noted the shadow of bamboos cast on the paper-covered windows by the moon. Her despondency fled before the beauty of the shadows. She traced the bamboo shadows on the paper window pane with a Chinese brush and black ink. The next morning she found the picture as attractive and appealling as when she traced it the previous evening. Other painters quickly adopted this idea, and the genre of bamboo painting became a standard subject in Chinese art. Love of bamboo, love of its rustling sound and fluttering shadow—all of these aspects have been incorporated into the role of bamboo in landscaping.

Left, bamboos, stone lamp and pergola. *Right,* a secluded bench in a bamboo grove at the Shanghai Botanic Garden.

THE EFFECT OF BAMBOOS IN LANDSCAPING

Bamboos can and should be arranged in a variety of ways to create diverse landscape effects. Here are some hints for garden designers.

DISTANT OR NEAR

Stoloniferous bamboos produce a splendid view from a distance. There is a rainy season in the Yangtze Valley during late spring, accompanied by many mizzling days. Sensitive and learned people greatly appreciate the undulating greenery bathed in misty drops, which enhances the green color and provides a clear yet vivid background. A bamboo grove with its fine texture is well suited to provide this background effect.

Clumped bamboos are best placed in the foreground with only a few clumps in the background. A few scattered in the middle ground to link the foreground to the background makes a harmonious and pleasing unity.

Bamboos and a 2-story hall in Chengdou.

A bamboo pavilion flanked by clumps of bamboo.

SCATTERED OR CROWDED.

Clumped bamboos should be scattered sparsely over a large area, or planted in tightly knit groups like trees—never evenly spread in the landscape. Even the stoloniferous bamboos can be confined in group plantings or divided by paths or streamlets. Such group plantings are very charming in a turfed area surrounding a lake or along a stream.

DENSE OR THIN.

A grove of stoloniferous bamboos or a single clump may be allowed to grow thickly for screening, or thinned out more or less severely to provide glimpses through the planting. Sometimes, just a few canes can make a fine picture against a lattice window, among several picturesque rocks, or beside a garden structure.

BRIGHT OR DARK.

Carefully thinning out some canes in a thick grove will make some small openings and allow the sunlight to filter into the darkness. The grove will be more interesting by having bright spots interwoven with dark spashes, especially in the early morning or late afternoon when the sunlight is slanting.

HOW TO USE BAMBOOS IN LANDSCAPING AND GARDEN MAKING.

It is fortunate that there are so many different species of bamboos. All the special requirements in garden making can be answered by this distinguished subfamily. Many species range to 20 meters high or more, while a few are less than one foot. The stoloniferous species can be used to make a tall hedge, while the clumped types can be treated as a standard tree. There are many ways to associate bamboos with other parts of a landscape and plants in a garden.

BACKGROUND OR WINDBREAK.

In old times, mandarins and rich people had their country villas built on south-facing sites and planted woods, orchards, or bamboo groves on the north side of the buildings to serve as a background as well as a windbreak. In Central China, where there is plenty of moisture, the stoloniferous species of bamboos are the preferred types.

If the bamboo grove happens to be planted on a small hill, the name 'Bamboo Hill' will be given the planting; if on the slope of a mountain, then 'Bamboo Slope'. It is most desirable to have large naked rocks or better a cliff at the back of the garden. If the site possesses these fine landscape features, they must be carefully left alone and not screened by bamboos, but rather with occasional bamboos sensitively placed to accentuate the hardness of the rocks and called 'Bamboo Rock' or 'Bamboo Cliff'.

Bamboo Groves

In the Jin Dynasty (317–419 A.D.), no one dared criticize the functioning of the imperial court, lest he face the executioner. Seven of the most famous literary people of the times secretly gathered to discuss Taoists' methods for becoming immortal and drink wine. To hide their activities they met in an opening deep inside a large bamboo grove and thus avoided imperial disfavor. Since that time they have been universally admired by the people as 'The Seven Sages Inside The Bamboo Grove'. During the same period, a squire and famous literary figure, Zhang Zhi, owned a large bamboo grove of over one hundred acres. In it he built a small house to escape the scrutiny of the repressive government. Once, the local governer, Xie Ling-Yun, who was also a famous poet, came to pay him a visit. Zhang hearing of his approach ran into the bamboos to escape a confrontation. He has been remembered through Chinese history for his courage in not bowing to a high-ranking official, with the epithet 'The Respectable Literary Man in Bamboo'.

As a result of these legendary occurrences, it became a common practice to build one's house, whether simple or elaborate, inside a bamboo grove in order to enjoy seclusion and tranquility. The simple pleasures of drinking tea or wine and

Bamboos flanking a memorial pavilion on Jianshan Islet near Zhenjiang. (Fong)

The tops of bamboo clumps appear above the wall of an inner court in the Shanghai Botanic Garden.

reading and writing poems with friends became the best entertainment. Such houses have a cozy name—'Nest Inside Bamboos'.

Not only in the suburbs, but also in the towns, some high-ranking officials built spacious gardens with large bamboo groves. In *Loyang Ming Yuan Ji* (Famous Gardens In Loyang), published in 1138 by Li Wen-Shu, three town gardens were noted as being distinguished by their magnificent bamboo plantings.

The first of these was the garden of Duke Fu-Zheng-Gong which had a huge bamboo grove in the northern part. Four openings, each ten feet wide, passed through the length of the grove, and were called 'Bamboo Caves'. One 'Cave' was oriented from east to west, while the other three ran from south to north. Streamlets ran parallel to each of the caves. Frequent by-passes and five pavilions were built inside the grove beside the caves so viewers might rest to contemplate the beauty of the garden and reflect on the garden as a microcosm of the Natural Order.

The second garden belonged to Marshall Miao, famous for its more than ten thousand giant bamboo canes.

The third was the West Garden owned by Sir Dong. Its central feature was a hall surrounded by bamboos. Before the hall, water sprang up from a stone fountain formed in the shape of a lotus. Windows on all four sides of the hall could be opened. It was cool inside the hall in summer, as the bamboos sheltered it from the sun. The author commented that you simply could not believe that such natural beauty existed inside a city.

The ancient Chinese landscape architects called paths passing through bamboo groves 'Bamboo Paths', while the streamlets which meandered through them 'Bamboo Streams'. In hot climates, a bamboo grove was undoubtedly the best air-conditioned spot before the introduction of mechanical systems.

Top, clumped giant bamboos and their shadows appear as huge baskets of Guilen. *Middle,* bamboo grove along the bank of a small river in Seven-Star Park, Guilin. *Bottom,* a large clump of giant bamboo can be very fascinating when associated with water.

BAMBOO AND WATER SCENERY.

One of the most highly regarded features in Chinese gardens is a water scene. It is ideal if the garden site lies beside a lake or on the bank of a river; otherwise, a man-made pond is built. The Chinese think a garden without a water feature looks like a face with the eyes covered. Water reflects, as do eyes, the floating clouds, the sun and the moon, passing birds, and the various objects near its shore. The Chinese poets have traditionally used the term 'eye wave' to explain why a casual glance from a pretty lady gives young men heartquake.

Bamboos appreciate humidity both in the air and in the soil. Therefore, 'riverside bamboos' and 'pondside bamboos' grow well as long as drainage is adequate. Bamboo is evergreen, so it decorates the water the year round. The water surface ripples, and bamboo twigs sway harmoniously when a breeze stirs. Both have a beauty that is quick-moving, lively and gay. If a bridge crosses the water, a few clumps of bamboos beside the bridge will give the created scene the polished appearance of a natural work of art. Just see with your mind's eye—moonlight, a running streamlet, a red-balustraded bridge, clumps of bamboos—what a picture! It will certainly set up in every Chinese intellectual a longing for his beloved and arouse sweet memories of the past, as the tradition of centuries suggests.

Bamboo—symbol of honesty; water—symbol of purity. Chinese poets have traditionally linked these qualities with bamboo in their poems. Yu Shinan (558–638), one of China's most renowned and respected calligraphers and poets, wrote the following on pondside bamboos:

The green tops brush the cloud adrift,
casting a shadow down on the clear pond.
The reflection flutters as ripples spread,
Flowing water keeps the twigs waving.
Dragons guard their secluded habitat,
Wings of Phoenix caress the water surface.
Wish to know their hardiness in nature,
Come to see at the end of the year.

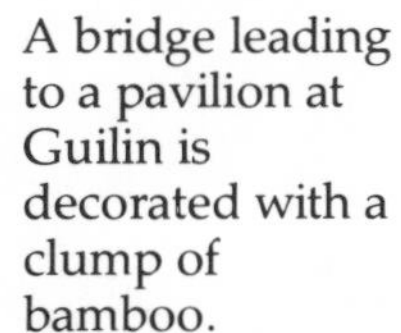

A bridge leading to a pavilion at Guilin is decorated with a clump of bamboo.

Twin bridges under a bamboo grove in Seven-Star Park, Guilin.

Bamboos and rockery against the wall of a small garden in the Shanghai Botanic Garden. (Fong)

BAMBOO AND ROCK.

Rock scenery is another important feature in Chinese gardens. The respected literary people and sages cherished the memory of the mountains and rivers in the places where they had been brought up or where they retired to study and reflect. When settled in town, carefully selected and sited rocks were installed in their gardens to bring back to mind their beloved mountains.

The emperor and painter, Song Hui Zong (1021–1086), ordered the removal of thousands of tons of rock from Taihu Lake to build a rockery in the capital Kaifeng, about 900 kilometers north. After the occupation of Kaifeng by the ruler of a Northern China Dynasty (1115—1234), some of these rocks were in turn transported to Beijing to enhance the royal garden there. Some of these rocks can still be seen decorating the piled-up Hill of Longevity in the famous Yi He Yuan, Imperial Garden.

Rocks in Chinese gardens are often arranged in one of three ways:

Dwarf bamboos grow from rocks which form a bank for the pond surrounding the pavilion in the background, Yangzhou. (Fong)

SOLITARY ROCK.

This is a single rock, skillfully and artfully placed so as to be the central focus of a broad open area. Rocks used for this purpose must be porous, with a furrowed surface and an unusual and fanciful shape. A few of the most splendid examples of such rocks have been named, and their history can be traced for hundreds of years. A few bamboos can be planted in a cunning relationship to a solitary rock to emphasize the rigidity of the rock and so create contrast.

ROCKPILE.

This type of rockery is built up with rocks only. They are piled together in a way designed to mimic natural cliffs and include caves and crevices. Bamboos and other plants are usually planted close to it, but seldom on it.

ROCKERY.

This type of rockery is a large man-made piled up hill made with rocks and earth. Huge rocks are scattered artistically on the sides and in some cases are placed at carefully selected points at the foot. Trees and shrubs, including bamboos, are used on a large scale for landscaping most of the hill so it looks like a mountain clothed in vegetation.

It is desirable to half-bury some large rocks between bamboo clumps and water to curb both and make the entire arrangement more attractive. Islands and islets are also ideal places to plant bamboo. Islets marked by flourishing bamboo clumps are a common sight in Chinese gardens.

The author has seen in a garden at Szuhou some bamboo canes springing up through a big hole in the center of a large rock laid flat inside a thin bamboo grove.

People plant bamboos beside rocks because the light, fluttering twigs make a distinct contrast to the massive rocks. The bulk and weight of rocks impart a feeling of oppression, but such scenes are softened by planting elegant plants like bamboo in relation to them.

Top, dwarf bamboos on lakeside rocks in Heyuan, Yangzhou. *Middle,* tall bamboos grow wild among the rocks in Rock Forest near Kunming. *Bottom,* bamboos and rocks surround a lawn in Rock Forest.

Opposite page

Top, bamboos seen through the plum flower-shaped moongate in the entrance of an old private garden in Shanghai. *Middle,* bamboos outside a modern trellised window and corridor at Suzhou. *Bottom,* bamboo clumps in an inner court, Chengdou.

BAMBOO IN THE INNER GARDENS OF PRIVATE RESIDENCES.

Private residences of high ranking mandarins or rich merchants frequently consisted of two to four rows of many-roomed buildings with open spaces between the rows. Each was separated into paved courts or inner gardens by walls, galleries, or occassionally a set of three-room side buildings. Although such large residences are no longer built, quite a number of houses with inner gardens still exist.

Each row of buildings is dominated by the Grand Hall located in the middle of the row. Usually such complexes have three rooms on each side of the Grand Hall. Typically, another 'Flower Hall' and a room are added at the end of each row. Larger residences would obviously add further rooms at each end.

The front court before the Grand Hall is usually paved, but plantings are provided for by containers in which shrubs, penjing (miniature trees & shrubs), and perennials are grown. The inner courts between the Flower Halls were developed into gardens.

When the author was in his teens, a friend occupied part of one such large residence. The inner garden was about 500 sq. m. and included a terrace, a pond, and a rockery with a small pavilion on the top; all carefully planted with trees, shrubs, and perennial plants.

Among the plant materials used in an inner garden, bamboo is quite popular. One of the most celebrated calligraphers of the Jin Dynasty, Wang Zi-You (?–388), upon moving to a new house, planted bamboos immediately. One of his friends asked him why he was in such a hurry to plant bamboo. He replied that he could not live there even one day if there were no bamboos.

It was a common practice to grow a few clumps of bamboos near stone railings and inlays of a terrace, or at the side of stone steps leading to the garden from the hall. The ancient Chinese poets and artists enjoyed lying on a chaise near the railing of a terrace where they could enjoy the cool breezes and be screened from the summer sun by bamboo. In such cool and peaceful conditions, they could concentrate upon difficult mental work.

Some people prefer to plant stoloniferous kinds of bamboos in a narrow border along the south wall opposite the hall. During the evening and night, the bamboos appear indistinctly and so make one feel more at ease and as if facing a whole grove. But the planting of bamboos should not be overdone in an inner garden, and they must always be curbed in such limited spaces.

There are usually a few small open spaces between buildings in these large residences which are difficult to landscape. For example, there commonly is a small area, less than 20 sq. m. behind the Flower Hall which is not large enough for a tall tree, but is a cozy corner for a dozen bamboo canes, as they require a very limited foothold, yet they are tall enough to reduce the intense sunlight of a summer afternoon. They can be seen through the north window of the Flower Hall, and the owner can 'lie by the north window, and enjoy the coolness', as suggested by a poet when asked where to take a nap on a long summer afternoon.

A clump of bamboo inside a typical moongate.

BAMBOO AND GARDEN STRUCTURE.

Strictly speaking, old Chinese gardens were architects' gardens, as landscapers played but a minor role. Both the spacious Palace Gardens in the North and the small private gardens in the Yangtze Valley were dominated by a series of buildings. The main garden was related to the principal buildings which are always located on the best part of the site. In an analogous manner, every minor structure is the center of a small garden scene. Indeed the standard work on garden-making, *Yuan Ye,* published in 1634, dealt mainly with architecture and treated horticulture as an adjunct to architecture and architectural features.

The modern gardens in China maintain this centuries-old tradition. They are designed to show what a public garden might be, but to the eyes of Western visitors, architectural structures dotted about public gardens seem obtrusive.

The traditional garden structures are:

***Tang* (Hall).** Tang is the main building in a garden, in which formal meetings, the reception of guests, literary activities, banquets or, before the liberation, gambling parties were held.

Tai. An elevated terrace with roof but usually having no walls except on the north. Partitions made of wood with carved trellis-work windows are fitted into the open sides on cold days. Tai is designed for the owner to enjoy distant views and to receive close friends or for literary gatherings which might end in drinking and feasting.

Ting. A pavilion, without walls and supported by pillars often elevated and approached by steps. Ting is used for resting or viewing. It is the most frequently encountered structure in gardens. It can be square, rectangular, hexagonal, octagonal, or even circular in shape. A half ting built against a wall can make a fine picture.

Ge. A two-storied pavilion with wooden walls. The upper story is used as a cozy place filled with the sweet smell of burning incense, for receiving intimate friends, making poems, drinking wine, and enjoying distant views.

Xie. A large rectangular roofed pavilion with wooden walls, located beside a water feature, and sometimes associated with a terrace. It is used to enjoy the sound and movement of water and to entertain.

Lang. A covered passage, open on one side as a gallery, or open on both sides. It is built to link two buildings, or when built along the front wall of a building, as a passageway. Garden scenes are carefully thought out to please the eye of viewers passing through.

Lou. Two-storied Tang or Xie serving the same functions as Tang.

The main purpose of all these structures is to provide a sheltered place for people to view the surrounding scenery. But on the other hand, these structures, as well as the people inside, become a part of the entire scene and can be enjoyed by people outside in the garden. Whether the structures are of a simple, rural type with straw-covered pavilions and bamboo buildings or of a more elaborate type using wood and stone, they demand the decoration of plants. However, the planting should not obscure the buildings around which they are grown.

Bamboos used for decorating garden structures should not be too tall, too thick, nor too close to the building, even if used as background, so that the elaborate building roofs and eaves are not blocked from view. For the same reason, pillars, carved trellis-work windows, balustrades, etc.

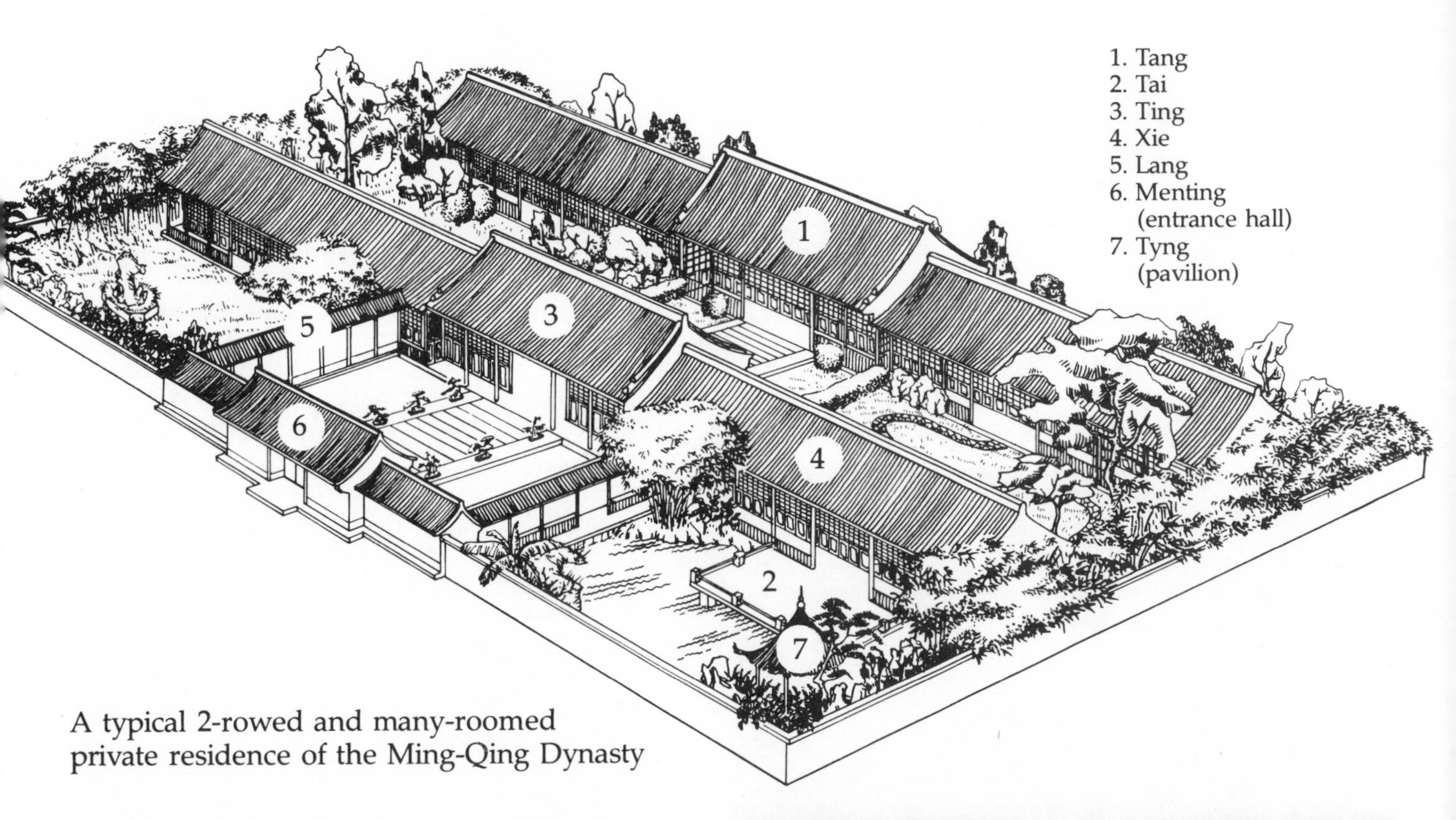

A typical 2-rowed and many-roomed private residence of the Ming-Qing Dynasty

should not be too heavily screened by plants. Excepting those buildings used as secluded studies, no garden structure should be surrounded with bamboos, even if they are thinly planted.

It is not a good practice to flank both sides of a building with bamboos, and it is worse to block the front with a large clump. A few canes, instead of a thick clump will make the building more attractive. In case of a long open gallery, a few clumps scattered here and there is much better than to plant a long continuous border. In such cases, some dwarf bamboos clipped into a neat hedge, as I have seen in the South China Botanical Garden, are very attractive but require much labour to maintain.

Sometimes it is wise to plant large clump bamboos before a Lou (two-storied pavilion or hall) or at the corner of it. The distant view can be more interesting if seen over the bamboo tips when standing on a balustraded balcony or leaning against the window sill of the upper floor.

The main gate of Geyuan in Yangzhou, famous in the old days for its bamboo landscape. Most of the bamboos have been recently replanted.

Bamboos alongside the corridor of a building in Yangzhou. (Fong)

Bamboos viewed through a rock-flanked moongate, Yangzhou.

BAMBOO AND OTHER GARDEN PLANTS

Bamboo is a highly prized plant for its hardiness and evergreen habit in the basins of the Yellow River and the Yangtze River, and especially when intermingled with Mei Flower (garden varieties of *Prunus mume*), grown for its fragrance and early-flowering habit. They are, together with *Pinus* spp., denominated the 'Three Friends in Deep Winter'. The Mei Flower comes into bloom before the leaves appear, and so needs bamboos as a background. In a like manner, a mature pine tree sets off the bamboos by its dominant role in the whole scene. When the 'Three Friends' are planted together, they form a traditional-picture. Just as Mei Flower represents the season of spring, so the other seasons are represented as follows: Chinese Cymbidium—summer; Chrysanthemum—autumn; and Bamboo—winter. The sensitive gardener will want to have all of them to form a traditional setting.

Bamboos are equally valuable when planted with other early-flowering trees, which bloom before their foliage develops, such as *Magnolia denudata, Malus halliana, Flowering Peach, Plum salicina,* and many others. Bamboo furnishes a fine textured background of evergreen foliage for all of them.

A garden without perennials such as herbaceous paeonies or chrysanthemums to give color splashes in late spring or autumn would be a poor and dull thing. A retired court official and poet, Wu Mei-Cun, in the early 17th century expressed a common human need in one of his poems. The first two lines are:

Untrodden mosses encircle my orange hedge and
thatched cottage.
Planted with my own hands are bamboos and perennials given freely
at my request.

He expressed in his poem that he was living on limited means, yet with clear conscience, but did not fail to make his garden with bamboos and perennials. Of course, some perennials should have their own beds, but others like *Hemerocallis, Hosta, Iris, Lycoris, Liriope, Rhodea,* and *Ophiopogon* grow well on the verge of bamboo borders if given careful attention.

The author wishes to quote a mocking poem by one of China's most famous poets, Su Dong-Pu (1037–1101), to conclude this chapter.

No bamboo makes one vulgar.
No pork makes one thin.
If you want to be both graceful and strong,
Have dish of pork and bamboo shoot every day.

Bamboo flanking rocks and wall in a small inner garden in Shanghai.

A pavilion covered with rice straw atop a hill and surrounded by bamboos near the water-side is an ideal place for viewing the beautiful scenery.

5 PENJING OF BAMBOO

Penjing means miniature landscape in a container. Whether the landscape is represented by a single tree or several as in the case of Tree Penjing (Bonsai in Japanese), or rocks with or without clinging plants as in the case of Rock Penjing, it certainly will remind you of the scenery which you have seen and admired in the wild. That is the reason the Chinese people, especially those living in town value Penjing so greatly. They remember enjoying the plants and rocks existing in Great Nature.

The history of Penjing can be traced back more than 1,200 years. The royal family of the Tang Dynasty enjoyed them greatly. Pictures of a court maid holding a Penjing with both hands were found in the tomb of a Tang prince. Among the various plants used in Penjing, bamboo is a traditional favorite. A small clump of dwarf bamboo, less than one meter high, in a small pot is as well balanced and meaningful as a twenty meter high, thick grove found in nature. If the Penjing-maker uses rocks as the main feature and puts bamboos at the foot of the miniature mountain or in small pockets up the cliff, the miniature landscape becomes more complex and requires a large container—some are up to several meters long, which is very impressive indeed.

HOW TO CHOOSE BAMBOOS FOR PENJING MAKING.

Bamboo has its own individuality as a Penjing plant and is appreciated especially by those of fine taste. Not all the species, only the dwarf ones with small leaves and thin canes, are suitable for penjing.

The best clumped species for this purpose are *Bambusa multiplex, B. multiplex* f. *alphonsokarri, B. multiplex* var. *nana, Bambusa ventricosa,* and *Sinocalamus affinis* var. *flavidorivens;* while the stoloniferous kinds are *Phyllostachys bambusoides* f. *tanakae, Ph. nigra, Ph. propinqua. Chimonobambusa quadragularis, Shibataea chinensis,* and those species of *Sasa* introduced from Japan.

HOW TO DWARF BAMBOOS.

Except for the *Sasa* species, the abovementioned bamboos normally reach a height of four to five meters when fully grown and so have to be dwarfed for making Penjing.

Select three small shoots in early spring and separate them from the mother plant; in the case of stoloniferous species keep the rhizomes intact. Plant them in a 12″ to 14″ pot. Use sandy soil with plenty of humus as potting material, and add some manure in the bottom of the pot. After potting, bury the pot in a sand bed up to its rim. Keep the soil moist but never waterlogged. Weak liquid fertilizer should be applied once a month after the plants are established. Do not shade them, save on hot and dry days. The plant will be much dwarfed and can be divided again when crowded. Divisions are usually required every two to three years.

The plants obtained from the divisions can in turn be potted-up in 5″ pots, or used directly for making Penjing. If a plant is still too tall to be appropriate for its place, the top can be cut off and side branches allowed to develop.

Nursery plants obtained from layering, cuttings or rhizome cuttings, potted into 5″ pots in the second year, make good Penjing subjects.

Penjing horticulturist Hu Rong-qing and his bamboo penjing. (Fong)

Dwarfed *Bambusa ventricosa* in a square pot 14 cm. deep.

Bamboos 60 cm. tall in a shallow round pot.

Small bamboo penjing in a 12 cm.-wide pot.

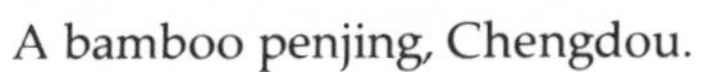

A bamboo penjing, Chengdou.

Dwarf *Sasa* with variegated leaves in a round pot 10 cm. in diameter.

Bamboos grow on the back of an elephant-shaped rock.

BAMBOO PENJING.

In round pots, only clumped kinds can be used. If stoloniferous bamboo are used, they will send up shoots around the edge and leave the center vacant. For larger rectangular containers, several clumps can be arranged just like a group planting of shrubs in a lawn corner, or stoloniferous kinds can be used to imitate a bamboo grove. If the root system of an available plant is too deep for the container, leave the upper part of rhizome unburied to make the Penjing look more ancient. Support is a necessity after planting. Thinning is rather important in Penjing composed of bamboos only. Never allow them to become too thick, and always remove the oldest culms when the young shoots grow up.

The smallest bamboo Penjing have to be placed in a sand bed under shade during hot weather. Syringe them several times a day. When potting, do not cover the drainage hole with crocks, but use a piece of strong leaf blade instead. When the soil ball is formed by the entangling roots, the leaf blade will have rotted allowing the moisture of the sand bed to rise up to the pot soil by capillary attraction. The plant gets extra water supply in this way, to its benefit.

Rocks added to penjing for decoration.

PENJING OF ROCK AND BAMBOO.

Just as in landscaping, the combination of rocks and bamboos is very attractive in Penjing. Bamboos can be planted not only at the foot of a big rock, but also in pockets of soil at the back, up the side, or even on the top. The pockets in the rock are carefully made by chiselling, filled with soil, and provided with adequate drainage. The whole miniature landscape might in time become too large for a container. It can then be put in a pond on an elevated stool without a container as an islet.

Other trees with colorful foliage or fruits can be used sparingly for decorating such a rock and bamboo Penjing. Some people like to put a few lead or stone pavillions, houses or sail boats in Penjing, while others reject the idea. Human figures are seldom used.

The maintenance of rock and bamboo Penjing is quite simple. Just pour water up to the rim of the shallow container and keep the pockets of soil moist by syringing. Occasionally, you have to remove a couple of old culms when crowded. A little fertilizer might be applied to the bamboo pockets from time to time, but never overdone.

If the container is not too large, it can be placed on a stand by the window in a sitting room or study, so you can enjoy the landscape whenever you are tired from reading too much.

Bamboos on a rock penjing at the Shanghai Botanic Garden.

A 7 m. tall rock penjing at the Shanghai Botanic Garden.

Bamboos and rocks are part of a very large penjing at the Shanghai Botanic Garden.

癸亥年
清秋月

6 HANDICRAFTS OF BAMBOO

As many of the articles used by the Chinese in everyday home life, including large furniture, were made of bamboo for thousands of years, Chinese craftsmen have acquired a deft skill and fine taste in working with it. In the past they made many splendid articles to please the aristocracy, landlords, and the very wealthy. Some of the work was so fine it was regarded as a form of art rather than a traditional craft.

The crafts of bamboo are now concentrated in the Yangtze Valley, and the craft can be classified into the following categories.

BAMBOO CARVING

Three different types of carving depend upon the source of the material carved. The first is Carving of Bamboo Culm Duckfoot. As duckfoot is nearly solid, it is carved in the same ways as conventional wood. Human figures and animals are the usual subjects.

A section of the culm internode with a node as bottom is used for the contemporary kind of carving called Carving of Culm Sections. Culm sections are mainly carved to provide pen and brush holders. The surface of the culm may be carved as in bas-relief. Alternatively the carving may pierce through the culm, hollow out all the blanks, and leave the design and figures free standing.

Thin bamboo skin is removed, smoothed out, and glued on the surfaces of small wooden boxes. The skin is designed and carved to show scenery, flowers, or human figures. This is called the Carving of Bamboo Skin.

BAMBOO WEAVING

Bamboo skin can be split into very thin, narrow, and long strips, which are ideal for weaving around armatures to make a wide variety of articles. Some of them, such as cases and baskets, are used both for ornamental and everyday purposes. Others are, however, used purely for ornamental purposes. They include figures of fairy tales or fables. (Armatures are made of split bamboo sticks.)

In the past, bamboo handicrafts were made of bamboo only, but now other materials are taking a minor part in bamboo handicraft. For example, a bamboo chandelier requires some silk tassels.

Two antithetical phrases carved on 60 cm.-long bamboo sections. Such plaques are often hung on the pillars that flank a door in a small garden structure. (Fong)

A bamboo lampshade. (Fong)

Bamboo brush holders. (Fong)

Surface carving on sections of bamboo culm with a node as the bottom. The specimen on the left is also pierced.

A pair of mandarin ducks woven from bamboo. (Fong)

Carved bamboo decoration on a bow with painted bamboo arrow shaft. (Fong)

Tea set with woven bamboo coverings. (Fong)

A bamboo basket. (Fong)

Below right, a bamboo birdcage. *Below left,* vase with elaborate woven bamboo covering. (Fong)

Bamboo furniture and decorations in a modern Shanghai hotel.

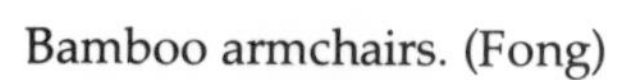

Bamboo armchairs. (Fong)

A bamboo partition. (Fong)

A bamboo bookcase. (Fong)

Bamboo-paneled pillar. (Fong)

梧竹幽居亭
癸亥仲秋
為大鈞先生著作寫生
鑄劍樓
曉钢

7 THE CULTURE AND PROPAGATION OF BAMBOO

There are differences of culture and propagation between stoloniferous bamboo and clumped bamboo, so it is better to deal with each type separately.

A STOLONIFEROUS BAMBOO.

They are, originally, subtropical plants, but many are quite hardy in the basin of the Yangtze River, with a minimum winter temperature of −10° to −15°C., while some can even stand a minimum temperature below −20°C. and so can be grown in regions of the Yellow River. Among the hardiest spp. are *Phyllostachys bambusoides* and *f. tanakae, Ph. propinqua, Ph. nigra, Sinarundinaria fangiana,* and *S. nitida,* but some shelter is required for them in very cold areas.

Stoloniferous bamboos grow vigorously and have stout underground root systems. They require fertile sandy or sandy loamy soil, at least 50 cm. deep, with good drainage. The pH value of the soil can range from 4.5 to 7. Alkaline soil or heavy clay are not recommended for bamboo.

The best time for planting is from November to February in mild climates, while it is safer in northern areas to wait till the danger of frost is past.

If there are no adequate materials available from a nursery, starts can be obtained by division from an established grove. It is important to select culms one to two years old together with stout rhizomes three to five years old. Each culm should have a piece of rhizome 70 to 100 cm. long. The culm base should be at a point approximately 4/10 the length of the rhizome, i.e., the 'coming rhizome' (from the rear end to the culm base) is no less than 30 to 40 cm. long, while the 'going rhizome' (from the culm base to the front end) 40 to 60 cm. long. The top of the culm should be cut off, with only 4 to 5 clusters of branches preserved. For dwarf or semidwarf species, it is advisable to dig up a piece of rhizome with 3 to 5 culms and plant them as a group for better results.

Do not plant too deeply, just 3 to 5 cm. deeper than the original soil level. After planting, mulching and staking are essential. Of course, abundant moisture should be given without delay.

NURSERY PRACTICE FOR PROPAGATING STOLONIFEROUS BAMBOO

1. Rhizome Planting. Select rhizomes 2 to 5 years old for tall-growing spp., and cut them into pieces 100 to 150 cm. long. Rhizomes of dwarf or semi-dwarf spp. should be 2 to 3 years old and 60 to 100 cm. long. Put two rhizomes flat in every planting hole, and cover them with 10 cm. of soil, which will be elevated slightly at ground level. Water well and mulch afterwards.

An improved method is to make the rhizomes shorter, about 50 cm. long. Lay them flat in a bed in March, and cover them with 4 to 6 cm. of soil. Keep the soil moist. About 1½ to 2 months later, new shoots will appear. Thin to 2 or 3 shoots per rhizome. Shade them during the first summer. They will be 60 cm. tall at the end of the first growing season. They may be removed to their permanent position after two years.

2. Rhizome Planting with Shortened Culms. Select 1 to 2 years old culms and cut them down to 15 to 20 cm. above ground level. Dig the shortened culms with rhizomes as long as recommended in the preceding paragraph. Better results are obtained in this way, but more material is required.

3. Seed Sowing. Seed can be sown in spring. It is reported that better results come from sowing seed by dibbling, 8 seeds for each hole, with a distance of 20–25 cm. apart. Seed germinates in about 3 weeks after sowing. Shade seedlings during the first summer. They may be planted in their permanent position after two to three years.

CLUMPED BAMBOO.

Most species are tender plants. However, some can stand a few degrees of frost and grow well in the basin of the Yangtze River, such as *Bambusa multiplex* and its variety *nana, Brachystachyum densiflorum, Pleioblastus gramineus, Pl. hindsii, Shibataea chinensis,* and *Sinocalamus affinis.* But they require some shelter in colder climates near the Yellow River.

The soil conditions good for stoloniferous bamboo are also suitable for clumped types. Their planting time should be a little later owing to their tender nature.

The best material for planting is strong culms one to two years old. Carefully keep the end of the culm (duckfoot) intact. Each culm should have two clusters of branches reserved on every axil below 1.5 to 2 m. Culms of big bamboos may be planted separately, while those of dwarf species should be planted in groups of 3 to 5 culms. They can be planted either in an erect or a slanted position.

NURSERY PRACTICE OF PROPAGATING CLUMPED BAMBOO

1. Trench Layerage. In March or April, select 1 to 2 year-old strong culms on the edge of the clump. Shorten all the main branches of the lower 20 axils to the second node, and remove all the side twigs. Make a cut through two-thirds the diameter of every culm just above the ground then bend them down for trench layering. Cover the lower 20 internodes in the trench with soil, and leave their tops uncovered. Cut the top off leaving only one axil above the ground, which should include all the twigs and leaves. Three months after layering, roots and shoots should appear from every axil. They can be separated the next spring.

2. Trench Cuttage. This procedure is the same as trench layerage, except the culm is separated entirely from the clump. The culm may have its duckfoot intact or cut off.

3. Single-node or Twin-node Cuttage. Select 2 to 3 year-old strong culms, and cut them to single node or twin node cuttings. The nodes below the midpoint of the culm provide single node cuttings, while the upper nodes provide twin node cuttings. Cuttings should be made 10 cm. above and 20 to 25 cm. below the node. Place the cuttings flat in a bed and cover them with 3 cm. soil, or strike the cuttings at an angle of 20 degrees from level and leave the top uncovered. It is best to cover the bed with polyethylene and also shade it from summer sun.

The germination of seed and the growth of seedlings.
Phyllostachys pubescens Mazel ex H. de Lehaie.

4. Branch Cuttage. Those species having a root-bud near the base of branch can be propagated by branch cuttage. Take 1 to 2 year-old branches with prominent root-buds from 2 to 3 year old culms. Cut the top of branch off at 2 cm. above the 3rd node and cut off all the side twigs except the third node with one twig and some leaves. Strike the cuttings at a 30 to 40 degree angle, leave the top twigs uncovered, and mulch the bed with 3 cm. of grass or straw. Shade the bed for the first ten days. Roots usually appear after 2 to 3 weeks.

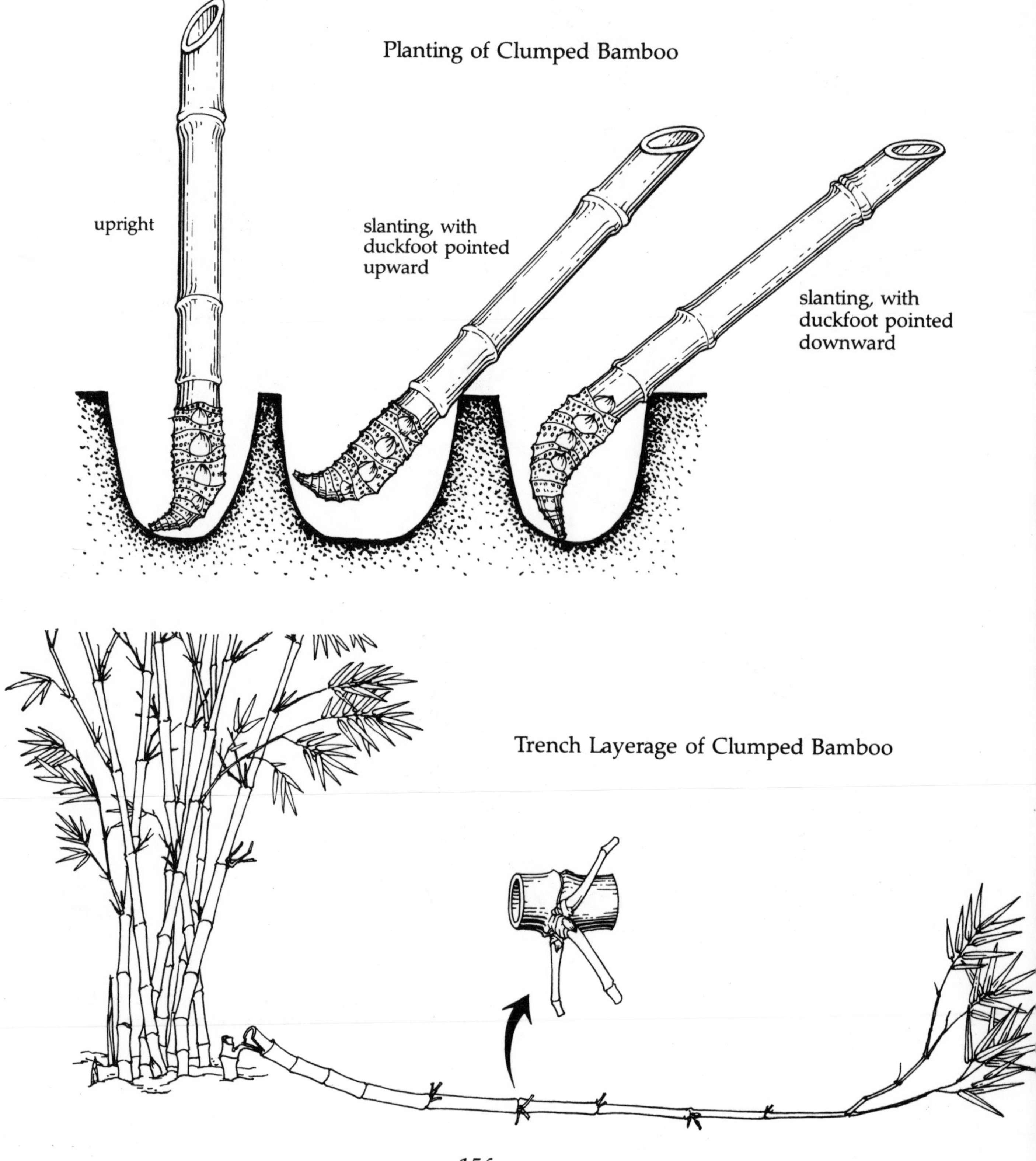

How to strike cuttings of clumped bamboo

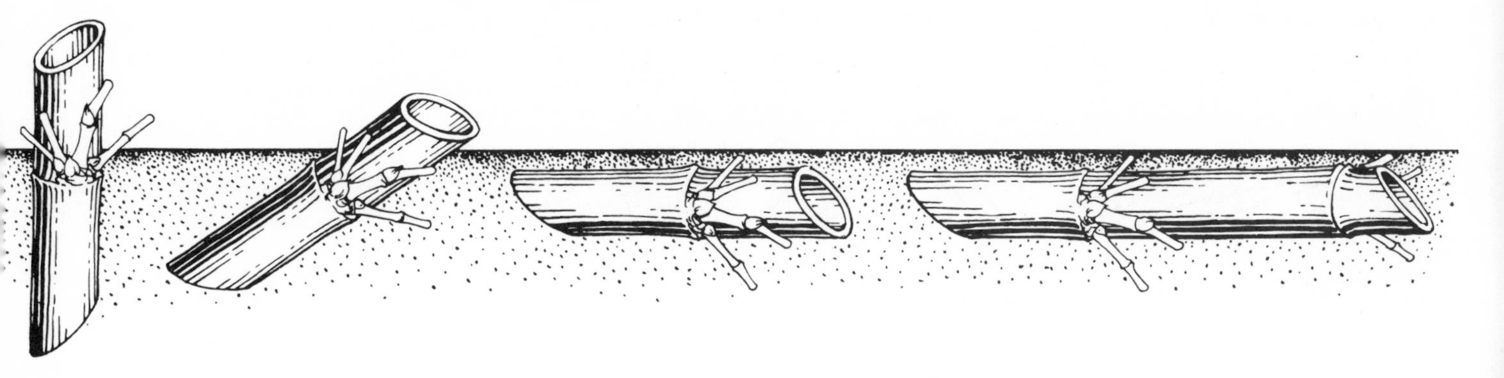

Cuttings of Clumped Bamboo

Qualified culm of *Phyllostachys pubescens* for transplanting, 3 to 6 cm. in diameter. First branching node is 1.5 m. or less from the ground. Length of coming rhizome is 30 to 40 cm., that of going rhizome 50 to 70 cm.

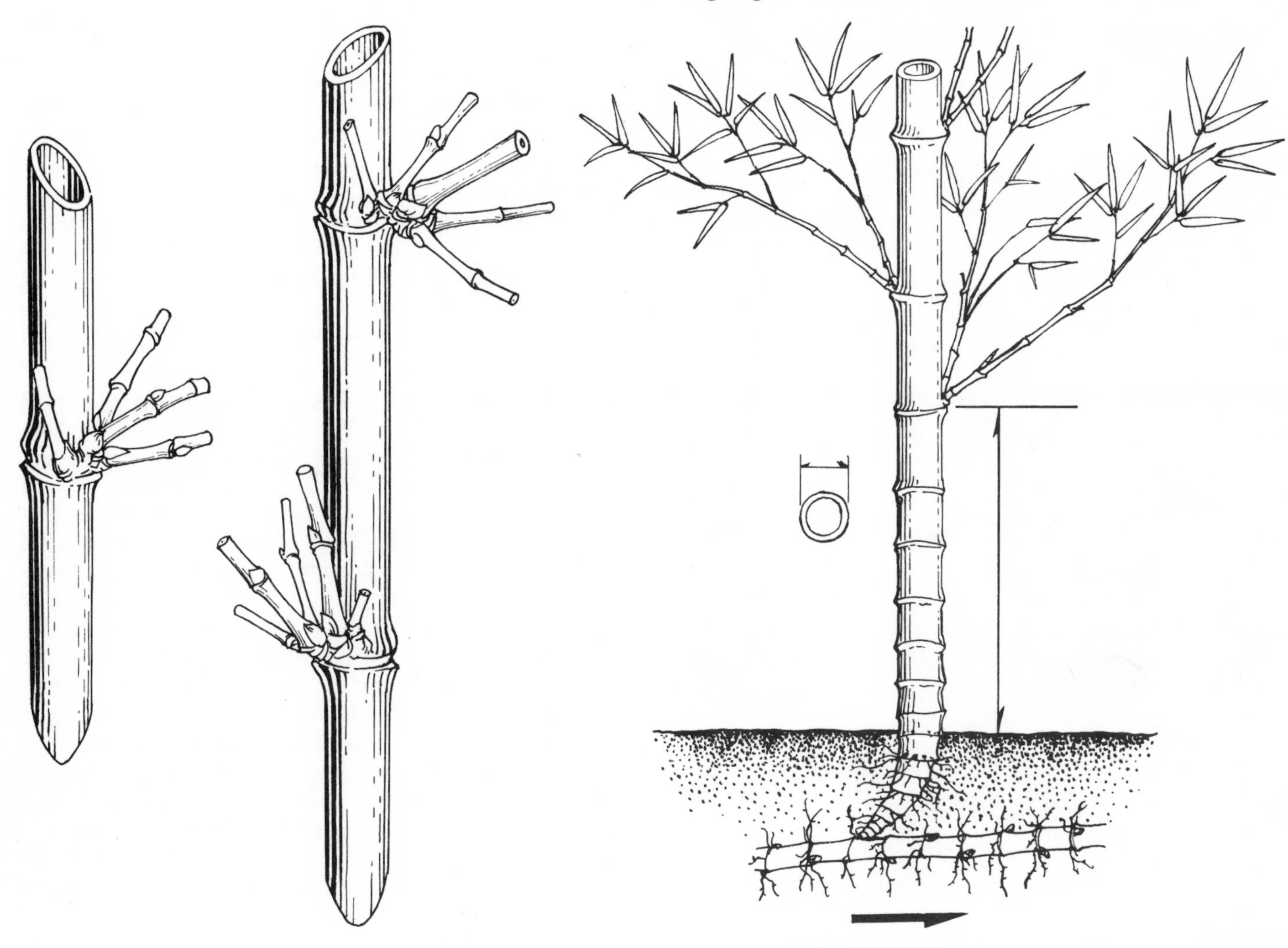

A nest in
bamboo
offers
secl

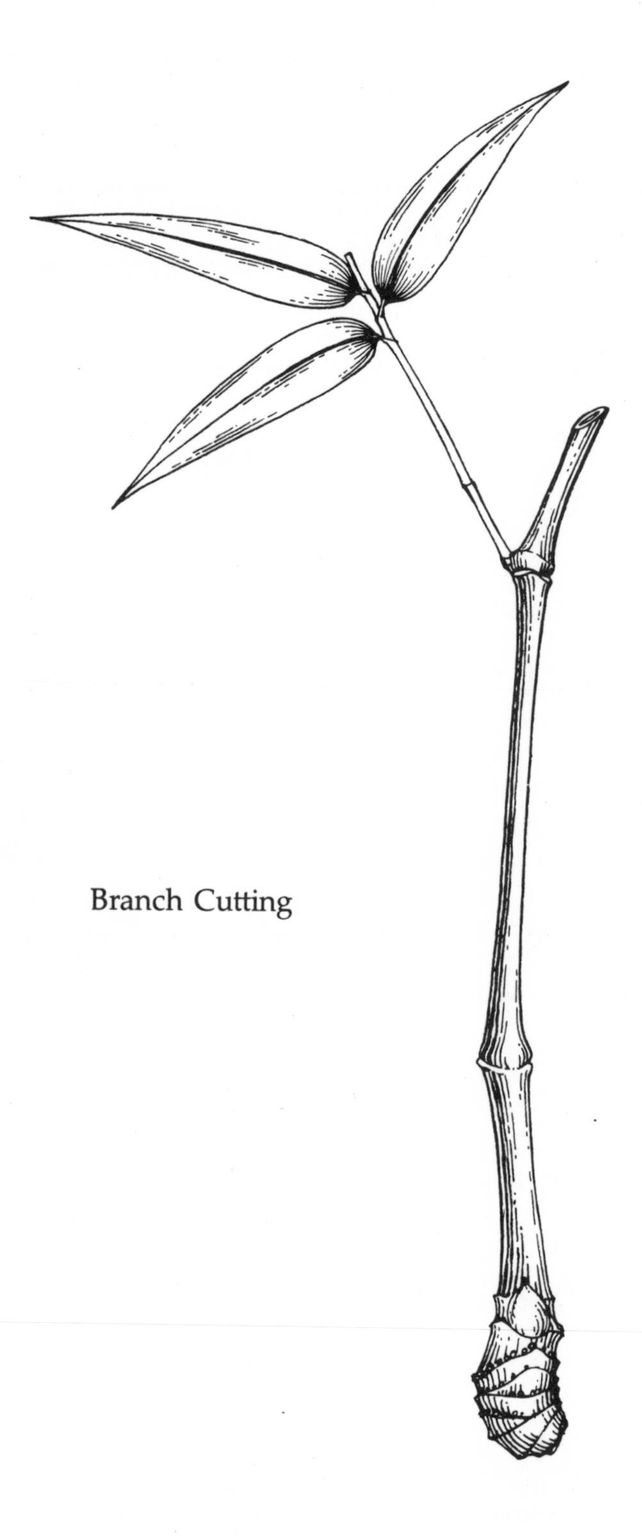

Branch Cutting

MAINTENANCE

Moisture. Bamboo likes high air humidity and requires regular watering in dry climates, especially during the spring and summer months. Ample water should be given immediately after transplanting. Soggy soil causes deterioration of the roots and should be corrected by providing adequate drainage.

Fertilization. Mulching with manure in winter benefits the growth of bamboo greatly. Chemical fertilizers may be applied during the spring and summer months. The best proportion of N:P:K is 5:1:7.

Pruning and Thinning. When a newly-planted bamboo plant sends forth too many shoots in the first growing season, allow only 2 to 3 shoots to develop into culms. Cut new culms off one-fourth to one-fifth of the whole length at the end of the season to promote the growth of roots. The top of new culms in an established clump or grove should never be cut off. If a clump becomes too crowded, old culms and weak growth should be thinned. Any stump in a grove of running bamboo should be carefully removed without hurting any healthy rhizome.

Delay or Avoidance of Flowering. Bamboo plants of most species will wither and die after flowering. All the plants in a large bamboo in the wild sometimes die in the same year. Such an event certainly causes a tremendous loss. But, fortunately, such disasters seldom occur in landscaping. However, when old culms show signs of flowering, they must be cut off immediately and large quantities of nitrogen must be applied to the new culms. In the rich soil of Zhijiang Province vigorous groves, now over 200 years old, have been maintained by carefully controlling flowering behavior.

Mountain, cliff, rapid stream, water-side pavilion and bamboo—a picture often found in the mountainous areas of southeast China.

8 CONTROL OF PESTS AND DISEASES

Pests and diseases can cause heavy losses in bamboo production areas, but they seldom damage bamboos in gardens. It is probable that the city environment and careful maintenance do not encourage them.

PESTS

LEAF-CHEWING INSECTS

Ceracris kiangsu Tsai is the worst of all pests. Sometimes the foliage of a whole grove might be killed and new culms killed as they sprout. The nymphs appear in mid-May, and must be controlled before they reach adult stage in early July by spraying stomach or contact poisons.

Caterpillars of *Algedomia coclesalis* Walker and *Artona funeralis* Butler do extensive damage to bamboo. Larvae of the former appear in June and July, while those of the latter appear three times a year in June, August, and October. Spray dipterex or DDVP for control.

SUCKING INSECTS

Maggots of *Harmolita phyllostachitis* Gahan. Eggs are deposited in holes bored in the top twigs. Maggots living inside the holes make the internodes shorter and thicker by sucking. Cut down and burn the infected culms.

Scale (*Chionaspis bambusae* Cockerell). Nymphs appear in June and July. Spray Malathion or Imidan for control.

Aphid (*Oregma bamsusicola* Takahashi). Aphids often transmit fungus disease (mainly *Meliola spp.*). Apply nicotine sulphate, Rotenone, or Malathion and cut down the fungus-infested culms, which should be burned at once.

BORERS

Larvae of *Purpuricenus temminckii* Guerin bore into culms over two to three years old.

Larva of *Cyrtotrachelus longimanus* Fab., *Otidognathus davidis* Fairmaire, maggots of *Pegomyia kiangsuensis* Fan, and caterpillars of *Atrachea vulgaris* Butler, all bore into bamboo shoots.

Culms and bamboo shoots damaged by borers should be dug out and burned.

DISEASES

Watery Withering. Cause unknown. Yellow water with a bad smell is found inside the internodes of culms withered in the summer months. Infected culms should be removed

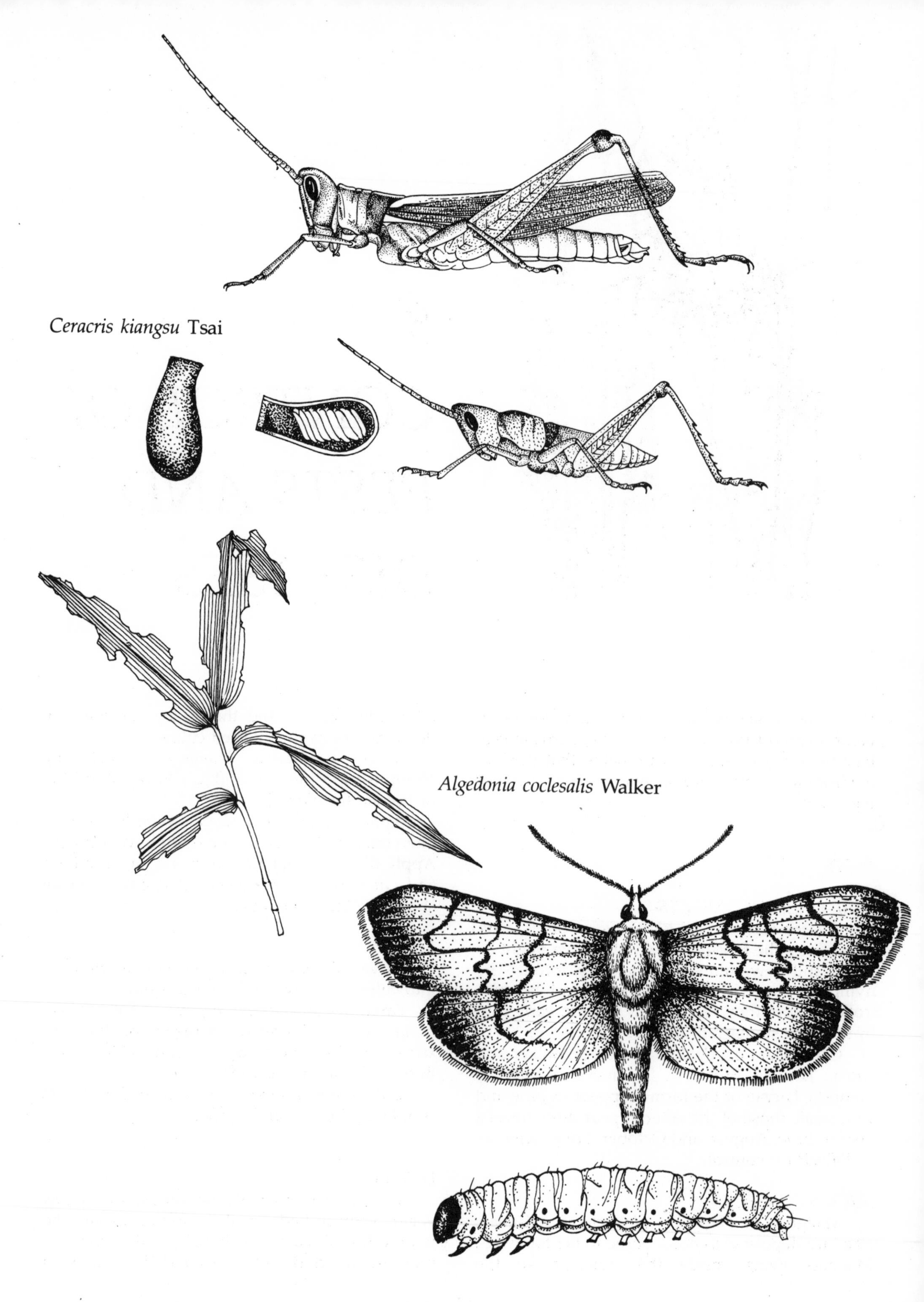
Ceracris kiangsu Tsai
Algedonia coclesalis Walker

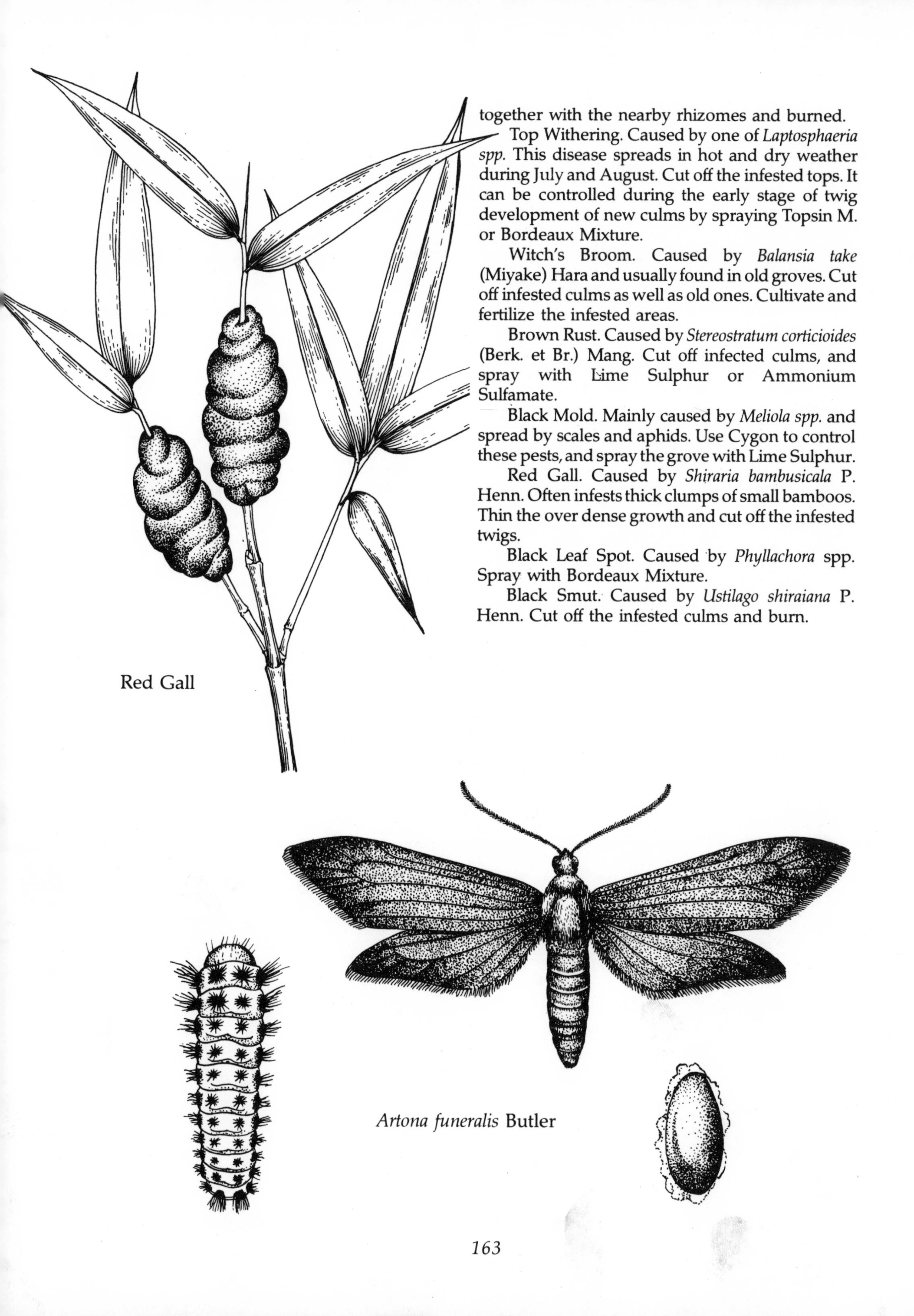

together with the nearby rhizomes and burned.

Top Withering. Caused by one of *Laptosphaeria spp.* This disease spreads in hot and dry weather during July and August. Cut off the infested tops. It can be controlled during the early stage of twig development of new culms by spraying Topsin M. or Bordeaux Mixture.

Witch's Broom. Caused by *Balansia take* (Miyake) Hara and usually found in old groves. Cut off infested culms as well as old ones. Cultivate and fertilize the infested areas.

Brown Rust. Caused by *Stereostratum corticioides* (Berk. et Br.) Mang. Cut off infected culms, and spray with Lime Sulphur or Ammonium Sulfamate.

Black Mold. Mainly caused by *Meliola spp.* and spread by scales and aphids. Use Cygon to control these pests, and spray the grove with Lime Sulphur.

Red Gall. Caused by *Shiraria bambusicala* P. Henn. Often infests thick clumps of small bamboos. Thin the over dense growth and cut off the infested twigs.

Black Leaf Spot. Caused by *Phyllachora* spp. Spray with Bordeaux Mixture.

Black Smut. Caused by *Ustilago shiraiana* P. Henn. Cut off the infested culms and burn.

Red Gall

Artona funeralis Butler

INDEX

Note: page numbers in regular (roman) type indicate general reference; bold type main reference; italic, illustration or (pl.) plate.